Biophysics
DeMYSTiFieD®

DeMYSTiFieD® Series

Biophysics
DeMYSTiFieD®

Daniel Goldfarb

New York   Chicago   San Francisco   Lisbon   London   Madrid   Mexico City
Milan   New Delhi   San Juan   Seoul   Singapore   Sydney   Toronto

The McGraw·Hill Companies

Cataloging-in-Publication Data is on file with the Library of Congress

McGraw-Hill books are available at special quantity discounts to use as premiums and sales promotions, or for use in corporate training programs. To contact a representative, please e-mail us at bulksales@mcgraw-hill.com.

Biophysics DeMYSTiFieD®

1 2 3 4 5 6 7 8 9 0   DOC/DOC   1 6 5 4 3 2 1 0

ISBN 978-0-07-163364-2
MHID 0-07-163364-2

Sponsoring Editor Judy Bass	**Project Manager** Manisha Singh, Glyph International	**Indexer** Robert Swanson
Editorial Supervisor Stephen M. Smith	**Copy Editor** Carol Loomis	**Cover Illustration** Lance Lekander
Production Supervisor Pamela A. Pelton	**Proofreader** Manish Tiwari, Glyph International	**Art Director, Cover** Jeff Weeks
Acquisitions Coordinator Michael Mulcahy		**Composition** Glyph International

For Sora Rivka, and for our children, Levi Yitzhok, Shaindel, and Menachem Mendel.

About the Author

Daniel Goldfarb has a Ph.D. in biophysics from the University of Virginia. He has done post-doctoral biophysics research, taught chemistry at Rutgers University, and written for the *Biophysical Journal*. Daniel is a Chartered Financial Analyst charterholder and currently applies his background in physics and math to designing and developing trading and risk analysis software for the financial industry.

Contents

Preface

Biophysics is a fascinating and relatively new science. After centuries of studying the physical properties and behavior of inanimate objects, we finally got the idea to use physics to study living things. The hope is to reveal the most basic principles of life, in much the same way that physics has illuminated fundamental principles of matter and energy.

This is an introductory textbook. Ideally we would cover the entire field of biophysics. You will certainly agree, as you read this book, that a thorough coverage of the entire field of biophysics is just not possible in a single volume. However, what you *can* gain from this book is a solid foundation, and a knowledge of biophysical principles and how they are applied. Once gained, that foundation will be easy to build on.

The book is organized as follows. The first three chapters give a broad overview of and introduction to the science of biophysics. In Chap. 1 you will learn to define biophysics, understand the prerequisites needed to study *Biophysics Demystified*®, and learn about the brief history of biophysics. You will also get to know the major divisions of biophysics and how the various topics in biophysics are categorized into these major divisions.

Chapters 2 and 3 then provide a broad overview of the science of biophysics. There are several reasons to gain a broad understanding of the field before getting into the details. First, you will need to build up some vocabulary, that is, the "language of biophysics." Understanding the most commonly used terms from the outset will be a help later on. Second, the various topics of biophysics are interconnected. Although each can be studied independently, a broad overview will give you the ability to understand the context of each topic as you learn its details. Similarly, knowing the interconnections will underscore the importance and usefulness of each branch of biophysics to the others. Finally, the broad overview of Chaps. 2 and 3 will enable us to cover a lot of topics that may not be covered in great detail later in the text, so at least you will understand how they fit into the whole. This will be part of your foundation for later learning.

Chapters 4 through 8 teach the principles of physics, biology, and chemistry that are necessary for a journey into biophysics. The focus is on aspects of these sciences that apply most directly to biophysics. This includes, in Chaps. 4 and 5, an understanding of free energy, the laws of thermodynamics, entropy, and statistical mechanics. Next, Chaps. 6 and 7 delve into the physical forces that come into play at the molecular level, again paying special attention to those that are most relevant to living things. We then review the major categories of biomolecules—what they are made of and aspects of their structure and function—sort of a quick overview of biochemistry from a biophysical point of view. Chapter 8 provides an overview of the living cell, its structures, and what these structures do.

In Chaps. 9 through 11 the focus is on subcellular biophysics. This is the most common and largest branch of biophysics, and so we go into it in more detail. This branch includes protein biophysics, DNA biophysics, and membrane bio-physics. Finally, Chap. 12 explores some aspects of anatomical biophysics, in-cluding blood flow and winged flight in animals.

You can use this book as a self-teaching guide, to lay the foundation for fur-ther study or just to satisfy your immediate curiosity. It can also be used as a classroom supplement, explaining and clarifying topics that are not as simple in other texts. My hope is that in reading this book you will truly find "hard stuff made easy."

Daniel Goldfarb

Acknowledgments

Thanks to Judy Bass, my editor at McGraw-Hill, for the opportunity to write this book, and for her advice and guidance throughout; to Dr. Kenneth Breslauer, a role-model biophysicist, for coming to my aid on short notice and helping me find a good technical reviewer; and to Pragati Sharma, for her great job in finding errors, and for her valuable suggestions for clarity in many places. I especially thank my wife, Sora Rivka, for her sound advice, her encouragement, helping schedule time for me to write, and everything else she did to help make this project possible. Finally, I thank my children for their exceptional curiosity, and for always being there (when I was trying to write).

Biophysics

DeMYSTiFieD®

chapter 1

Introduction

What Is Biophysics?

Plain and simple, biophysics is the physics of biology, just as astrophysics is the physics of astronomy and nuclear physics is the physics of atomic nuclei.

What does this mean? What *is* the physics of biology? Physics is the study of matter and energy. Biophysics tries to understand how the laws of matter and energy are at work in living systems. Another way to say this is that biophysics uses of the principles, theories, and methods of physics to understand biology.

Biophysics is an interdisciplinary science. One could say it is the place where physics, chemistry, biology, and mathematics all meet. In practice, most biophysicists study things at the molecular level, but biophysics also includes physiological, anatomical, and even environmental approaches to the physics of living things.

Prerequisites for Biophysics

Biophysics is an advanced science. It requires some basic knowledge of biology, physics, chemistry, and mathematics. However, since this book is meant as an introduction to biophysics, to *demystify* biophysics, then as much as possible along the way, I will introduce or review the necessary background information.

Still, I do need to make some assumption as to your level of understanding in the physical sciences. For this purpose I will assume that you have had at least an introductory college-level (or advanced high school–level) course in physics or chemistry. This may be self-taught. Notice the *or* in *physics or chemistry*. If you have one or the other, it should be enough to get you through this book.

Let's see two examples. For our first example, take the equation

$$F = ma \tag{1-1}$$

This should look somewhat familiar to you. It means that if you apply a force F to an object of mass m, you will cause that object to accelerate with a rate of acceleration a.

The equation also says that for any given force F on an object of mass m, the acceleration rate will be exactly the rate that causes the product (mass × acceleration) to be equal to the force. This means, if we replace the object with a more massive object so that m is larger and we apply the same force, then the acceleration must be smaller (so that the mass times the acceleration will still be equal to the force). Similarly, applying the same force to a less massive object will cause the object to accelerate faster.

Let's put this in concrete terms. Say we apply a force of 12 newtons (N) to an object with a mass of 2 kilograms (kg). The object will accelerate at a rate of 6 meters per second per second (or 6 m/s²). That is, every second, the object will be going 6 meters per second (or 6 m/s) faster than it was the previous second, for as long as we continue to apply that force. If the object is standing still when we first apply the force, then after 1 s, it will be moving at 6 m/s, after 2 s, it will be moving 12 m/s, and after 3 s, 18 m/s [about 40 miles/hour (mi/h)].

But, if we now apply that same force, 12 N, to a more massive object, say 24 kg, that object will accelerate much more slowly, only $^1/_2$ m/s² (see Fig. 1-1).

To take this example a step further, we write Eq. (1-1) as

$$\mathbf{F} = m\mathbf{a} \tag{1-2}$$

Notice that the **F** and **a** are now bold. This means that they are vectors. A *vector* is a quantity that has not only a size but a direction as well. A force is always

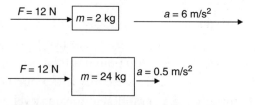

FIGURE 1-1 • If we apply the same force to two objects of different mass, the more massive object accelerates slower. The product, force = mass × acceleration.

applied in a specific direction. Acceleration also occurs in a specific direction, and the acceleration will occur in the same direction as the force. The mass of the object, however, is not a vector quantity, but is called a *scalar*: a quantity that has only size.

All of this should sound familiar to you. By now you should at least be thinking, "Yes, I remember that. It's all coming back to me." (Unless you're already thinking, "I know this. When's he going to get to the biophysics? I knew I should have skipped this section.")

Let's take one more example. Consider the chemical reaction

$$CO_2 + H_2O \rightarrow sugar + O_2 \qquad (1\text{-}3)$$

The concept of a chemical reaction and this way of representing it needs to be, if not familiar, at least something you can grasp quickly and become comfortable with. Equation (1-3) means that carbon dioxide and water can react together to form sugar and oxygen. This is one of the most basic biological reactions. It is how plants capture carbon and energy from the environment and store that energy in a form (sugar) that we (and other animals) can later consume and use to stay alive and to go about our daily business.

Strictly speaking, this reaction should have been written with a double arrow, like this:

$$CO_2 + H_2O \leftrightarrow sugar + O_2 \qquad (1\text{-}4)$$

The double arrow means that the reaction can occur in both directions. Going from left to right, carbon dioxide and water combine to create sugar and oxygen. This direction requires the input of energy. Plants get that energy from sunlight and, through the process of photosynthesis, store some of that solar energy in the chemical bonds of the sugar.

The reverse direction, from right to left, releases energy through the oxidation or combustion of sugar. In living things this process is also known as respiration. It is the way living things release stored energy that can be used for moving around, growing, and so on.

To emphasize the fact that energy is required to combine the carbon dioxide and water and that energy is released when the sugar is oxidized, we can also show energy as part of the chemical reaction.

$$CO_2 + H_2O + energy \leftrightarrow sugar + O_2 \qquad (1.5)$$

The preceding discussions of forces and chemical reactions should be something that you can feel comfortable with.

Still Struggling

If you lack the necessary prerequisites to be familiar with or to feel comfortable with the preceding discussion, then I highly recommend you begin with *Physics Demystified* by Stan Gibilisco. Alternatively (or additionally) you may prefer *Chemistry Demystified* by Linda Williams. If you have a strong preference to start with physics or chemistry, then by all means go with your natural inclination.

A Brief History of Biophysics

How old is biophysics? As a separate discipline, biophysics is a relatively new science. In the big scheme of things, it is far, far younger than physics, mathematics, chemistry, or biology, but somewhat older than genetic engineering or computer science.

Although we find some physical studies of living things scattered throughout history, biophysics as a discipline is only about 60 to 100 years old. The first published use of the word *biophysics* was in 1892, in *The Grammar of Science* by Karl Pearson. In the book, Pearson tells us that there is a need for a new scientific discipline. In Pearson's words, "The reader might conceive that our classification [of the sciences] is now completed, but there still remains a branch of science to which it is necessary to refer." He explains that there appears to be no link between the physical and biological sciences and points out that "a branch of science is therefore needed dealing with the application of the laws of inorganic phenomena, or Physics, to the development of organic forms." He proposes that the new branch of science be called *bio-physics*.

Although Pearson coined the term, I like to mark the birth of biophysics with the series of lectures given by Erwin Schrödinger in 1943. Schrödinger won the 1933 Nobel Prize in Physics for his work on quantum mechanics. In the 1930s a small handful of physicists began turning their attention to the questions of biology and biochemistry. Then in February 1943, Schrödinger gave his now famous lecture series titled "What Is Life?" The Friday afternoon lectures were so popular they had to be repeated on Monday for those unable to fit into the lecture hall. A year later the lecture series was published as the book, *What Is Life? The Physical Aspects of the Living Cell.*

The lecture series and book had a major impact on several notable scientists of the time. Only a few years later, in 1946, the Medical Research Council of King's College in London established the Biophysics Research Unit of King's College. Their goal was to hire physicists and put them to work on questions of biological significance. The physicist Maurice Wilkins and the physical chemist Rosalind Franklin were among those who joined the unit to become biophysicists. There at King's College they used X-ray diffraction to investigate the structure of DNA.

The particle physicist Francis Crick at Cambridge University was also inspired to turn his attention to biophysics. He was soon joined by the biologist James Watson. In 1953 Watson and Crick made one of the most far-reaching discoveries of our time when they used Rosalind Franklin's X-ray diffraction data to discover the double helix structure of DNA.

In 1957 the Biophysical Society was founded, to encourage growth and dissemination of knowledge in biophysics. Since then, interest in biophysics has only increased. By the early 1980s numerous universities offered graduate degrees in biophysics, but only a few, if any, colleges offered undergraduate degrees. Today over 60 colleges and universities offer undergraduate degrees in biophysics, and the Biophysical Society has over 9000 members. Figure 1-2 shows a time line of science and biophysics to put the age of biophysics in perspective.

The Scope and Topics of Biophysics

Biophysics is a very broad science, including a wide range of activities such as

- Studying the forces between atoms that determine the shape of a protein or DNA molecule
- Developing algorithms for a computer to analyze and display a three-dimensional image of the brain in real time during brain surgery
- Investigating and comparing the mechanics of limb movements or blood flow in various organisms
- Researching the effects of radioactivity on the environment

There are many ways we can classify the long list of topics that make up the field of biophysics. One very convenient and typical way to organize the broad

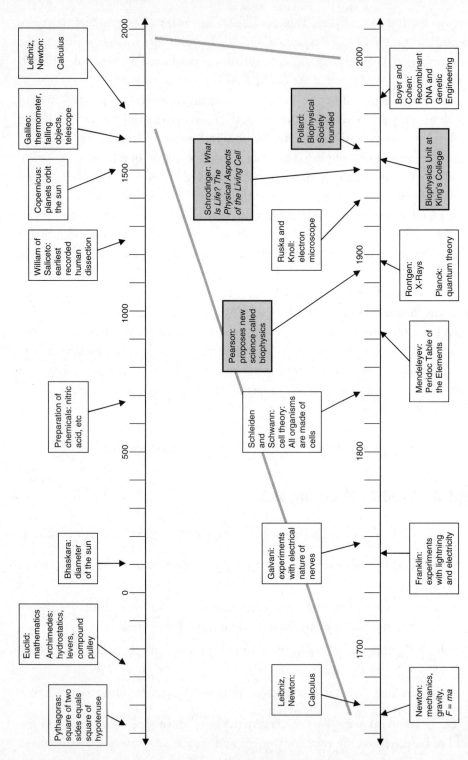

FIGURE 1-2 • How old is biophysics?

6

scope of biophysics is according to the relative size of what we're studying. For example, are we studying molecules, cells, or whole organisms? Another common and useful way is according to technique employed and application. With this in mind, and with some overlap, we will classify the many topics of biophysics into two broad classifications subdivided up into six categories.

- Biophysical topics based on relative size of subject

  1. Molecular and subcellular biophysics
  2. Physiological and anatomical biophysics
  3. Environmental biophysics

- Biophysical techniques and applications

  4. General biophysical techniques
  5. Imaging biophysics
  6. Medical biophysics

The next two chapters briefly describe many of the topics found in biophysics, organized according to the two major classifications just given. The purpose is to give you a broad overview of the scope of biophysics and to introduce vocabulary specific to biophysics. This will aid in understanding the detailed chapters that follow.

Throughout the book, vocabulary words that are important for you to learn will be in *italics* when first defined and will appear in the glossary in the back of the book.

QUIZ

Refer to the text in this chapter if necessary. Answers are in the back of the book.

1. An object of mass 3 kg is not moving. It is then pushed with a constant force of 12 N, causing it to accelerate at a rate of 4 m/s². After 5 s how fast will the object be going?
   A. 12 m/s
   B. 15 m/s
   C. 20 m/s
   D. 60 m/s

2. An average-size adult cheetah (50 kg) can accelerate from a standstill to 30 m/s in just 3 s. If we approximate the forward thrust provided by the cheetah's leg muscles as a constant force, what force is necessary to cause this acceleration (10 m/s²)?
   A. 50 N
   B. 150 N
   C. 300 N
   D. 500 N

3. If a cheetah accelerates from a standstill at a rate of 10 m/s² and reaches its top speed in just 3 s, how far will the cheetah have traveled in that time?
   A. 45 m
   B. 30 m
   C. 15 m
   D. 10 m

FIGURE 1-3 · The cheetah is the fastest land animal. An average adult cheetah weighs about 110 pounds (lb). Its powerful leg muscles generate enough forward thrust to accelerate the cheetah from standing still to a top speed of about 70 mi/h in just 3 s.

4. **A major source of energy for animals is carbohydrates (sugars) from plants (grains, fruits, vegetables, etc.). Where does this energy primarily come from?**

    A. Plants use their roots to draw energy from the ground.

    B. Potential energy was stored in the seed before the plant grew.

    C. Plants breathe oxygen at night and carbon dioxide during the day.

    D. Photosynthesis extracts energy from sunlight and stores it in chemical bonds.

5. **A chemical equation is a symbolic representation of a chemical reaction. A double arrow in a chemical equation indicates**

    A. we are not certain about the chemical reaction.

    B. the chemical equation is balanced.

    C. the reaction involves both physics and chemistry.

    D. the reaction can occur in either direction.

6. **Although Watson and Crick are credited with the 1953 discovery of the double-helical structure of DNA, they were only able to do this with data obtained by Rosalind Franklin at the Biophysical Research Unit of King's College in London. What kind of data did she obtain?**

    A. Temperature measurements of DNA crystals

    B. Electron microscopic images

    C. X-ray diffraction

    D. Magnetic resonance

7. **Biophysics is an interdisciplinary science; this means that**

    A. the internal aspects of living things are studied.

    B. it takes a lot of discipline to be a biophysicist.

    C. biophysics is less rigorous than other sciences such as physics and biology.

    D. biophysics combines physics, chemistry, and biology into a single science.

8. **Two common and convenient ways to classify the various branches of biophysics are by**

    A. size of what is studied and by technique utilized.

    B. size of what is studied and by whether mathematics is used.

    C. technique used and by application.

    D. technique used and by percentage of physics, chemistry, biology, and mathematics used.

chapter 2

Biophysical Topics

In this chapter, we present an overview of the various topics of biophysics according to the following three major divisions: molecular and subcellular biophysics, physiological and anatomical biophysics, and environmental biophysics.

CHAPTER OBJECTIVES

In this chapter, you will

- Gain an understanding of the broad scope of biophysics.
- Acquire some vocabulary of biophysics, learning the most commonly used terms and how they apply to the branches of biophysics.
- Learn to classify the topics of biophysics into the three major divisions of
    - Molecular and subcellular biophysics.
    - Physiological and anatomical biophysics.
    - Environmental biophysics.
- Learn how the topic areas of biophysics are interrelated.

Just a reminder: Throughout the book, vocabulary words that are important for you to learn will be in *italics* when first defined and can also be found in the glossary in the back of the book.

Molecular and Subcellular Biophysics

By far the most common branches of biophysics are those dealing with molecules and subcellular function. This division of biophysics is sometimes also called *biochemical physics, physical biochemistry,* or *biophysical chemistry.* All three terms mean the same thing—what we will call *molecular and subcellular biophysics.* It is the place where biology, chemistry, and physics all meet. Within this division of biophysics we find the following topics.

The Structure and Conformation of Biological Molecules

This branch of biophysics deals with determining the structure, size, and shape of biological molecules.

Many biological molecules are polymers. A *polymer* is a large molecule made by connecting together many smaller molecules. Each of the smaller molecules is called a *residue* (because that's what's left over when you break a polymer into pieces). The residues making up a polymer may be identical, like links in a typical chain where the links are all ovals. The residues may also be a set of related but not identical molecules—imagine a chain where the links are various shapes: circles, triangles, squares, and rectangles.

Biopolymers (biological polymers) often fall into the latter case, where the residues have something in common but are not identical. For example, proteins are made by linking together smaller molecules called *amino acids.* The details of what an amino acid is are not important right now. For now, all you need to know is that each of the residues making up a protein is an amino acid and there are about 20 or so different amino acids found in proteins. Various amounts of these 20 or so amino acids can be linked together in various sequences to make different proteins, just as the 26 letters of the alphabet can be put together in various amounts and various sequences to form different words and sentences.

There are *four levels of structure* in biological molecules: primary, secondary, tertiary, and quaternary.

Primary structure specifies the atoms or groups of atoms making up a molecule and the order in which they are connected to one another. In polymers, rather than describe the primary structure in terms of specific atoms, we typically indicate only which residues we find and in what order we find them.

Secondary structure refers to the initial, simple, three-dimensional structure of a molecule. For example, a molecule, or part of a molecule, may take the shape of a helix or a shape similar to a pleated sheet.

Tertiary structure refers to the fact that a secondary structure, such as a helix or pleated sheet, can fold back on itself (sometimes over and over) and form a globular shape. As an analogy, if we consider an inflated balloon to be the

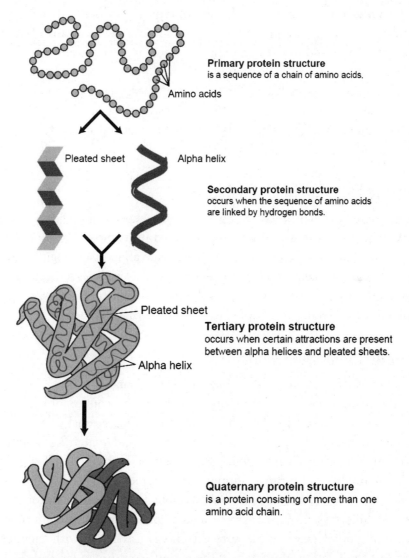

Primary protein structure
is a sequence of a chain of amino acids.

Amino acids

Pleated sheet   Alpha helix

Secondary protein structure
occurs when the sequence of amino acids
are linked by hydrogen bonds.

Pleated sheet

Tertiary protein structure
occurs when certain attractions are present
between alpha helices and pleated sheets.

Alpha helix

Quaternary protein structure
is a protein consisting of more than one
amino acid chain.

FIGURE 2-1 · The four levels of structure in biological molecules are illustrated here using the example of a protein, but apply as well to other molecules and sub-cellular complexes. (*Courtesy of National Institutes of Health: National Human Genome Research Institute.*)

three-dimensional secondary structure, then tertiary structure is folding and twisting that balloon into a balloon animal or some other creative shape.

Quaternary structure refers to the case where two or more tertiary shapes attach to one another to form an even larger molecule or complex. Extenting our balloon analogy, quaternary structure refers to using more than one balloon to make our balloon animal.

Not all biomolecules exhibit all four levels of structure. Small molecules (for example, simple sugars or amino acids) typically exhibit only primary and secondary structures. Biopolymers most commonly exhibit all levels up to tertiary structure, and sometimes exhibit quaternary structure.

The structure and conformation of biological molecules, as a branch of biophysics, also includes analyzing the forces and energy required for a molecule to maintain a particular shape. With this information, biophysicists develop geometric and mathematical models to predict the secondary and tertiary structure of a molecule, given its primary structure.

Structure Function Relationships

Closely related to determining the structure and shape of biomolecules is determining which parts of a molecule are involved in its biological function, and determining how changes to its structure or shape affect its biological function. When a one particular part of a molecule or complex is involved in carrying out its function, that part is referred to as the *active site* of the molecule. It is also possible for a molecule or complex to have more than one active site.

Conformational Transitions

Conformational transition is just a fancy term for a change in shape. Although the word *conformation* can mean structure *or* shape, in the context of biophysics it almost always means shape, specifically the three-dimensional arrangement of atoms in a molecule (that is, the secondary, tertiary, and quaternary structures).

Biomolecules often change their shape as part of their function. For example, the DNA double helix must temporarily unwind in order for the genetic instructions to be read or in order for the DNA to replicate itself for the next generation. Biophysicists use a variety of techniques to measure conformational changes in biomolecules, to measure the energy associated with them and to determine the relationship between the various conformations and their biological function. It is also possible to induce conformational changes in the laboratory. These induced changes may or may not happen in nature. In either case, induced conformational transitions can further our understanding of the forces involved and can be used to develop medical treatments and diagnostics.

Ligand Binding and Intermolecular Binding

A very common theme in subcellular biological function is the binding together of molecules. Sometimes the molecules are roughly equal in size and bind together to form a larger *complex* (quaternary structure). Each individual molecule in the complex is called a *subunit*. An example is hemoglobin, a large complex protein that carries oxygen from our lungs, through our blood, to the cells in our body. Hemoglobin is made up of four subunit proteins that bind together.

In other cases of molecular binding, a smaller molecule binds to a larger molecule. In such cases we call the smaller molecule a ligand. A *ligand* is a smaller molecule or atom that binds to a larger molecule. The smaller molecule may be integral to the biological purpose of the larger molecule, or it may simply serve to activate or deactivate the larger molecule in carrying out its purpose. When hemoglobin carries oxygen from the lungs to all of the cells in our body, oxygen molecules bind to the hemoglobin in the lungs. Later, the oxygen is released from the hemoglobin, so it can be used inside our body's cells. When oxygen binds to hemoglobin, the oxygen is considered a ligand.

You should be aware, however, that *sometimes* the word *ligand* may be used in any case of two or more molecules binding together (not just a smaller molecule binding to a larger one). In this book we will use the term *ligand* to mean specifically the case where a smaller molecular (or atom) binds to a much larger molecule. We will use the more generic terms *molecular binding, subunit binding,* or simply *binding* to refer to cases where the size difference between the two molecules is not significant (see Fig. 2-2).

Biophysicists studying ligand binding and other intermolecular binding seek to measure and understand

- The forces and energy involved in binding
- The interaction between multiple binding sites
- How changes to the molecules affect binding
- The relationship between binding and conformational transitions
- The relationship between binding and biological function
- The competition between different ligands that can bind to the same molecule
- The rate at which binding occurs and the factors that affect binding rates

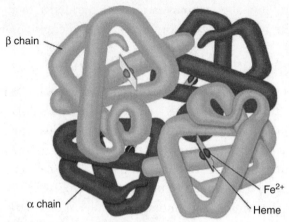

β chain

α chain

Fe²⁺

Heme

FIGURE 2-2 • Hemoglobin is a complex of four subunits. The four subunits consist of two identical pairs of proteins. One protein is called the alpha chain and the other is called the beta chain. The hemoglobin complex consists of two alpha chains and two beta chains. *Molecular binding* holds these four subunits together. *Ligand binding* occurs when oxygen binds to the hemoglobin. Each of the four subunits contains a group of atoms called heme. And each heme contains an iron atom (Fe^{2+}). An oxygen molecule, acting as a *ligand*, binds to the iron atom within each subunit. In this way, one hemoglobin complex can bind up to four oxygen molecules. (*Reprinted with permission of www.themedicalbiochemistrypage.org.*)

Diffusion and Molecular Transport

This branch of biophysics studies how molecules move around within cells and how molecules move from outside a cell to inside the cell and vice versa. In fluids, molecules are continually moving, randomly colliding, and jostling about. *Diffusion* is the process of molecules spreading out, as a result of this random motion. By spreading out, we mean that the random motion will cause molecules to move from a region of higher concentration (where they are closer together) to one of lower concentration (where they are further apart). The physics of diffusion can be described mathematically and can be used to better understand and predict biological activity in cells. Diffusion is the primary means of molecules moving around within a cell. However, as we shall see, living systems also have several other means of moving molecules to where they are needed.

Membrane Biophysics

All living things are made up of cells. The *membrane* is what defines the boundary between a cell and the outside world. Internally (inside the cell) membranes

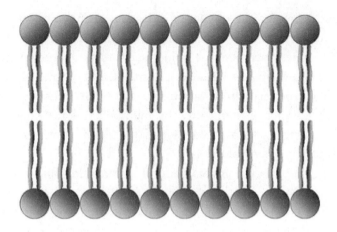

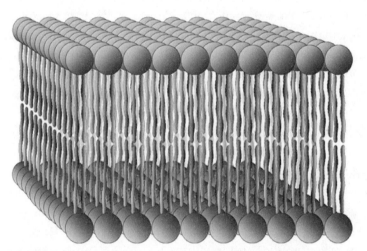

FIGURE 2-3 • Lipid molecules can stick together forming bilayers. These lipid bilayers are the main ingredient of cell membranes.

define and separate various parts of the cell from one another. Cell membranes are typically made of a double layer of lipid molecules. *Lipids* are fats or oils.

Lipid molecules have a string-like shape with a "head" at one end. The shape and physical characteristics of lipid molecules make them associate with each other (stick together) in the form of a *bilayer* (two layers) with the molecule heads on the outside of the bilayer and the string-like tails on the inside (see Fig. 2-3).

Membranes limit and control the movement of molecules into and out of the cell and from one region of the cell to another. Membranes are also able to create electrical potential across their surface, by controlling the flow of ions into and out of the cell. Understanding the physics of lipids and membranes can help us to better understand and predict how cells will behave under various conditions.

Membrane biophysicists often use lipid vesicles to study membranes. A vesicle is a small hollow sac. *Lipid vesicles* are small hollow spheres of artificial membrane that can be made from various types of lipids. Thus a lipid vesicle is like a cell with nothing inside it, just the membrane alone. This provides a simple tool to conduct experiments on the behavior of membranes without the complications of other parts of the cell.

It is possible to also place drugs or chemical agents inside lipid vesicles. Additionally, there are ways to attach certain molecules to the outer surfaces of such lipid vesicles to help the vesicles bind to specific sites in the body. In this way we can create *targeted delivery systems* to deliver drugs or chemicals to a specific location in the body (for example, to the site of a tumor). By understanding the physics of lipid conformational transitions, we can control these conformational transitions, and thus control the ability of the lipid vesicles to contain the drug or chemical inside them. Once the drug-filled lipid vesicles are in the bloodstream, we can apply a stimulus like heat or mild radiation to a specific part of the body to cause the lipid vesicles to release the drugs at that place.

DNA and Nucleic Acid Biophysics

DNA (deoxyribonucleic acid) is the biochemical that makes up our genes and controls our physical heredity. A closely related nucleic acid is *RNA* (ribonucleic acid), which serves many purposes within the cell. This branch of biophysics studies the physics of DNA and RNA. The secondary structure of DNA is a double helix, like two spiral staircases wrapped around each other. The double helix itself can bend and twist to form a helix as well. This helix of a helix is called a *superhelix*. The process of forming a superhelix in DNA is known as *supercoiling*. Supercoiling of the double helix is a tertiary structure in DNA. A quaternary structure in DNA occurs when the DNA superhelix wraps itself around protein complexes known as *histones* (see Fig. 2-4).

DNA biophysics includes studying

- Conformational transitions in DNA, including winding, unwinding, bending, stretching, and supercoiling
- Binding of proteins, RNA, and other molecules to DNA
- Energy changes associated with conformational transitions and binding and their effect on function

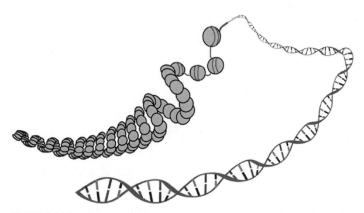

FIGURE 2-4 • The DNA double helix (secondary structure) bends and twists to form a helix of a helix, or superhelix (tertiary structure), which wraps around protein complexes known as histones (quaternary structure). (*Adapted from Wikimedia Commons.*)

Protein Biophysics

Proteins are involved in nearly every biological process within the cell. Examples include catalyzing biochemical reactions, regulating (turning on and off) biochemical processes, and transporting molecules across cell membranes, from cell to cell and from one part of a cell to another. Proteins are also involved in cell motility (self-induced movement of the cell). In order to carry out these functions, proteins typically must fold into very specific shapes, bind with other molecules, or undergo one or more conformational transitions. Since proteins do all of these things as part of their normal function, understanding the physics of protein folding, conformational transitions, and binding is crucial to understanding and possibly controlling their role in biological processes.

Bioenergetics

This branch of biophysics studies the physics of energy flow in living systems. Bioenergetics is concerned with all levels and branches of biophysics, from the environment, to the organism, to the cell, and to the molecules within the cell. At the core of bioenergetics is the study of how organisms and cells obtain the energy they need to carry out biological processes. This includes where the energy comes from, how the energy is stored, how the energy is converted into various forms, and where and how excess or unusable energy is released. While every branch of biophysics needs to be concerned with energy, some biophysicists specialize in understanding the energetics of any biological process, whether the process is protein folding, DNA unwinding, respiration, or energy flow in the environment.

Thermodynamics

Very closely related to bioenergetics is the study of thermodynamics. The laws of *thermodynamics* describe how energy behaves in physical systems, biological or otherwise. The *first law of thermodynamics* states that energy cannot be created or destroyed. The *second law of thermodynamics* states that in a closed system the orderliness of the system can never increase, but can only decrease over time.

At first glance, because living things are so complex and highly organized and because they have the ability to stay organized, it would appear that living things may somehow violate the laws of thermodynamics, particularly the second law. But living things are not closed systems. They interact with their environment.

Yet as recently as the 1940s many scientists continued to consider the possibility that living things do not behave according to the laws of physics, at least as we know them. In Schrödinger's *What Is Life? The Physical Aspects of the Living Cell*, he speculated that we may yet discover new laws of physics at work in living things that are not apparent in the inorganic world. This is certainly an understandable speculation given Schrödinger's own work on quantum mechanics— work that showed that physics at the atomic and subatomic level is very different from the physics of everyday experience. However, decades of exhaustive thermodynamic and physical studies of living things only confirm that organisms follow the same laws of physics found in the nonliving universe.

Statistical Mechanics

Statistical mechanics is the application of probability and statistics to large populations of molecules. Although it is impossible to measure the exact energy or state of every one of the trillion billion molecules in a test tube or cell, it *is* possible to develop models of how those molecules behave mechanically. A *model* in our case is a mathematical description of how the molecules move, how much energy they have, how they change shape, and so on. The model is then used to calculate the statistical probability of an event, for example, the probability of a protein molecule undergoing a shape change needed for its function.

Once the probabilities are known, they can be used to calculate statistical averages for the entire sample (that is, for the entire population of molecules in the test tube or cell). These statistical averages, in turn, can be associated with specific things that we *can* measure. For example, the statistical averages can be used to calculate and predict thermodynamic quantities such as temperature, pressure, and amount of energy released or absorbed. In this way, even though it is impossible to directly measure what each and every molecule is doing, *statistical mechanics allows us to interpret* the things we can measure in terms of what specific molecules are doing.

The interpretation is not direct knowledge, but we can design experiments so that the results either support or disprove our interpretation of what the molecules are doing. This is an important point in biophysics and in science in general. Obviously we want experiments to agree with our ideas of how the physical universe behaves. But it's even more important to design experiments that attempt to prove that our model is wrong! If we design an experiment to disprove our model and it fails to do so, this is a stronger support of the model than an experiment designed to agree with the model.

Good experimental design allows us to consider a model "correct" in the sense that the model accurately predicts the results of future experiments and can be used as a tool to manipulate living things and biomolecules as we choose.

Kinetics

This branch of biophysics deals with measuring the rate or speed of biological processes such as biochemical reactions, conformational transitions, and binding or unbinding of biomolecules. Kinetics is closely related to energetics and thermodynamics. Thermodynamics tells us whether a given process or biochemical reaction will occur.

Kinetics tells us how fast it will occur. What's the connection?

For now, let's just say that a process will happen spontaneously if that process results in a system going from higher energy to lower energy. We learn this about a process by studying its thermodynamics. Think of a ball rolling down a hill. The ball has higher potential energy at the top of the hill and moves to a state of lower potential energy at the bottom of the hill. So the process of a ball rolling down a hill is spontaneous.

However, the rate at which a process occurs is related to the energy *path* of a process. That is, does the energy decrease gradually or does it drop quickly? Does the energy only decrease throughout the process, or does it decrease and increase and then decrease (perhaps multiple times) during the process? How fast a ball rolls down a hill, for example, depends on (1) how steep the hill is, (2) whether there are any increases in steepness or flattening out along the way, (3) the presence, height, and slope of any speed bumps, and (4) any other obstacles along the way. These factors all affect the ball's path, and through the path, they affect how fast it will roll down the hill.

If a process goes through intermediate steps along the way from its start to its end and one or more intermediate steps are higher in energy (for example, rolling up a speed bump), these are called *high-energy intermediates*. The presence of high-energy intermediates, like speed bumps, tends to slow a process down. One example of this is DNA replication, discussed in Chapter 10. In order to replicate, the DNA double helix needs to unwind temporarily. This is a higher energy state for the DNA. Once it is replicated, the DNA goes back to its double helical state, which is a lower energy state. Thus, the unwound double helix is the high-energy intermediate in the process of DNA replication.

Living systems often regulate their biological processes by modulating the *rate* at which they happen. This means that, sometimes, when a living thing needs a process to stop (for example, it already has manufactured enough of a certain type of protein and for the time being it doesn't need anymore), then instead of actually stopping the process it simply allows the process to slow down almost to a halt. The process is still happening, but at such a slow rate that it doesn't make much of a difference. Then, when the organism needs the process to continue (for example, it now needs more of the protein), it simply speeds up the rate at which the process is happening.

When an organism needs to modulate the rate at which a process happens, typically this is done in one of two ways: either by providing the energy needed to get over a speed bump (high-energy intermediate) or by providing an alternative path (effectively removing or going around the speed bump). Sometimes a faster path (one without a speed bump) is provided by a conformational transition, by ligand binding, or by the binding of proteins acting as catalysts. Taking the example again of DNA unwinding, the unwound DNA is a high-energy intermediate, like a speed bump. A protein called helicase binds to the DNA and unwinds the double helix. In part, the binding energy contributes to the energy needed to attain the high-energy intermediate (the unwound DNA).

Still Struggling

To better understand how an energy path affects the rate of a process, consider walking over a large hill versus walking around it. The end result is the same, you are on the other side of the hill, but the path is different. You might think going over a hill is faster than going around it, especially if you've got enough energy to easily hike over the hill, since going over a hill is often shorter than going around it. However, biological processes typically involve hundreds or thousands of molecules. For example, when a cell needs to manufacture a certain protein, it connects together hundreds of smaller molecules to make the protein. In addition to this, the cell typically needs to make hundreds or even thousands of copies of the protein. So to make our analogy complete, let's say 1000 people need to hike over the hill (or around it). The time it takes to get all 1000 people over the hill can be very slow if only some of the people have enough energy to hike over the hill. (The others are tired and hungry.) Let's also say there is a tall fence alongside the path, so the only way to the other side is to go over the hill. We have two choices to speed up the process of getting 1000 people to the other side of the hill. We can either provide the energy they need to get over the hill (carry them, or feed them), or we can remove the fence (a conformational change) and provide an alternative path around the hill. In a cell, molecules are all jostling about with various amounts of energy. If a biophysical process involves an energy path via a high-energy intermediate, then the process will proceed slowly if only some of the molecules have enough energy to reach the intermediate state. In order to speed up the process, the organism either must somehow provide the energy to get more molecules to the high-energy intermediate, or it must provide an alternative path to the end of the process (a path that does not involve a high-energy intermediate) so that even the low-energy molecules can participate.

Biophysicists studying kinetics also develop models to describe the molecular mechanism of a process. This is similar to the use of models in statistical thermodynamics, we have discussed. The model is a hypothesis as to what the molecular mechanism is that causes some process. Each possible model implies a particular energy path for the process. The model may also suggest or imply a means for changing the reaction rate, by modifying various conditions of the experiment in a way that would alter some part of the mechanism in a known way. Experiments measuring the rates of biological processes under various conditions can then either support or refute the model. If the rate is successfully altered as the model suggests, this lends support to the correctness of the model. If not, then the model is proven wrong, and we need to come up with a better model to explain the process. In this way we can arrive at a molecular interpretation of a biological process by measuring the rate of that process.

Molecular Machines

A *machine* is a device that can alter the direction and/or size of a force (for example, a pulley, a lever, or an inclined plane). A *motor* is a special type of machine that also has the ability to convert potential energy into mechanical energy, that is, into a mechanical force or motion. So the difference between an ordinary machine and a motor is that, in the case of a motor, the force being changed does not come from outside the machine, but is generated by the machine (the motor) itself. The motor can continue to generate mechanical force as long as it has a source of potential energy or *fuel* needed to do so.

Living things are full of machines and motors. For example, our muscles use our bones as levers to redirect and in some cases magnify or decrease the forces they apply. Muscles themselves are motors; muscle fibers have the ability to convert chemical potential energy from the food we eat into the mechanical force of muscle contraction. Other examples, besides muscle contraction, of organisms generating a mechanical force or motion include

- *Cilia.* These are hairlike projections on the surface of some cells that move, allowing the cell to swim. Or in the case of cells that don't swim but have cilia, the motion of cilia can be used to push substances past the cells. For example, cilia on the inner surface cells of the lungs help clean the lungs by moving dust and other particles up and out of the lungs.

- *Flagella.* These are longer, whiplike structures that stick out from the body of some cells and move to propel the cell forward.

- *Pseudopodia.* Some cells move by temporarily pushing out on their membrane at one or more locations, changing the shape of the cell and causing the cell to crawl along.

- *Secretions.* Cells that manufacture proteins or other substances to be used elsewhere in the body (for example, pancreatic cells that manufacture insulin) must somehow move the manufactured molecules from inside each of the cells where they are made, to the cell's surface, and through the cell membrane into the bloodstream. Secretory cells have various mechanisms for packaging and moving the substances they make.

- *Separation of chromosomes (DNA) during cell division.* When a cell is getting ready to divide, it first duplicates its chromosomes. The cell then has to somehow move and separate the two copies of the chromosomes off to two opposite sides of the cell so that the two daughter cells each get a single copy of the cell's DNA.

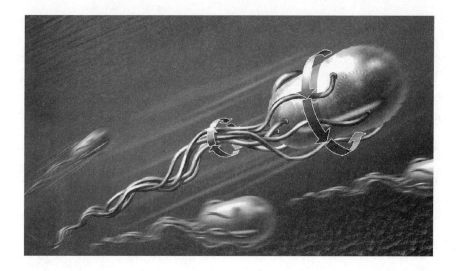

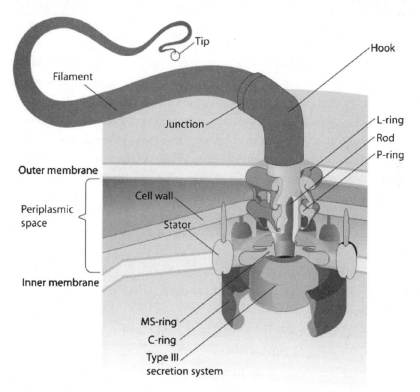

FIGURE 2-5 • Bacteria use flagella to swim. Flagella are whiplike projections that propel the cell forward. Each flagellum is a molecular machine embedded in the cell membrane. (*Courtesy of Wikimedia Commons.*)

In all cases where organisms generate motion, a close inspection of the source of that motion always comes down to *individual molecules acting as motors and machines* in order to generate and direct forces. For example, the motion generated by muscle contraction, at the lowest level, results from individual molecules of one protein (myosin) binding to and pushing against the molecules of another protein (actin).

Allosterics

Often we find that binding in one part of a molecule affects activity in another part of the same molecule. This property is known as *allostery* (from the Greek *allos* meaning "other" and *stereos* meaning "object" or "solid object"). Allostery allows two things to take place: allosteric regulation and cooperativity.

The control of a biological process by a cell or organism is called *regulation*. In many biochemical processes, one particular part of a protein molecule is directly involved in carrying out that protein's function (see also the section Structure Function Relationships). That part is known as the *active site* of the protein. Many biological processes involve binding or some other interaction between molecules at the active site. Regulation of the process usually happens simply by controlling binding directly at the molecule's active site.

Sometimes regulation of a process is achieved through binding somewhere other than the active site. This other binding site is called the *allosteric site* because of its long-distance effect on the active site. When a biochemical process is controlled in this way, we say that the process exhibits *allosteric regulation*.

Sometimes some sites behave as active sites and as allosteric sites at the same time, each affecting the other. The result is *cooperativity*, the occurrence of separate events together in a nonindependent manner. There are different degrees of cooperativity. In a slightly cooperative process, events occur only slightly more together than they would if they were completely independent. In a highly cooperative process, a set of otherwise independent events occurs in a mostly all-or-none manner.

The classic case of allosteric cooperative binding is that of oxygen binding to hemoglobin. Hemoglobin carries oxygen from our lungs through our blood to cells throughout our body. One hemoglobin complex can bind four oxygen molecules. The four sites that actively bind oxygen also act as allosteric sites, each enhancing the binding of oxygen to the remaining sites. The result is that four oxygen molecules tend to bind to hemoglobin in an almost all-or-none manner.

The long-distance effects of an allosteric site on another part of a molecule can be the result of conformational changes or of changes in intramolecular forces within the molecule. Allosteric processes typically show kinetics (changes in rate as a response to changes in conditions) that are different from the kinetics of nonallosteric processes. Biophysicists studying allosterics need to bring together several branches of biophysics, including structure, structure-function relationships, conformational transitions, ligand and intermolecular binding, and kinetics.

Physiological and Anatomical Biophysics

Biomechanics

In physics, *mechanics* deals with the application of forces to physical objects. The objects can be solids or fluids. For example, understanding how air flows around an airplane wing involves mechanics, as does understanding the path of a baseball when struck with a bat.

Biomechanics is the branch of biophysics that deals with the application of forces to biological objects. Biomechanics includes studying such things as

- *The motion of animals*. How animals apply forces to move. For example, how muscles and bones work together in walking, running, jumping, lifting, throwing, catching, and so on
- *The mechanics of various systems within the body*. For example, how air is drawn into the lungs, and how blood is pumped through the heart and throughout the body
- *Mechanics at a cellular and subcellular level*. For example, how cells move molecules around in their interior, and how cells respond to external forces

Biomechanics is obviously a very broad subject, having considerable overlap with many areas of biophysics. Notice also that biomechanics is studied at all levels: subcellular, physiological, and environmental. Over the years biomechanics has developed into somewhat of a discipline of its own (for which we could write Biomechanics Demystified). Thus, to some people, biomechanics is not so much a branch of biophysics but an engineering discipline geared toward the application of practical results. In this sense, biomechanics may be more closely related to bioengineering or biomedical engineering than biophysics. In practice, however, there are scientists who call themselves bioengineers, and scientists who call themselves biophysicists, and both use and study biomechanics. But there is a tendency that the field of biomechanics is more commonly found in the purview of the engineers.

Electrophysiology

Electrophysiology is the study of electrical aspects of living things. A primary focus of this branch of biophysics is the study of nerves. More generally though, electrophysiology is concerned with *excitable tissue*, that is, cell types that create, conduct, or use electrical impulses. Excitable tissues include nerves, muscles, sensory cells, and electrogenic and electroreceptive cells.

Sensory cells are specialized cells that are able to convert energy from some outside stimulus into an electrical response. For example, rod cells and cone cells of the eye contain proteins that can absorb light. The energy from the absorbed light causes a conformational change in the protein, allowing it to react with other molecules in the cell. This in turn starts a chain of events, resulting in an electrical charge on the surface of the cell. Similarly hair cells in the ear contain tiny hairlike projections on their surface. Sound vibrations in the air create pressure that pushes against these hairs, moving them. The movement causes these cells also to generate an electrical charge on their surface. An electrical charge on the surface of sensory cells typically stimulates electrical impulses in neighboring nerve cells. The end result is a sensory perception: we see light, we hear sound, we feel pressure, and so on.

Electrogenic and electroreceptive cells are found in certain electric fish such as eels and rays. Electroreceptive cells are specialized sensory cells that create a nerve impulse in response to an electric field. Some fish use electroreceptive cells to navigate. Electrogenic cells are similar to muscle cells. However, instead of generating contraction or movement, electrogenic cells build up strong electric charges on their surface. Electric fish use these strong electric charges for hunting and self-defense.

Electrophysiology includes

- Studying the workings of excitable tissue. For example, understanding the mechanisms of contraction in muscle cells, or exploring how various conditions affect the speed and strength of electrical impulses in nerves
- Understanding the electrical nature of the heart, and how to measure and regulate a steady pumping rhythm
- Studying, at a molecular level, how cell membranes of excitable tissues create electrical charges on their surfaces and propagate electrical impulses along the length of the cell
- Cable theory, using mathematics to model conduction of electrical impulses along nerves and along the detailed networks of nerve cells
- Investigating the mechanisms of various diseases that affect the ability of nerves or muscles to work

Sensory Biophysics

Sensory biophysics explores the electrophysiology and mechanics of the senses: seeing, hearing, touch, balance, smell, and taste. Take vision for example. Sensory biophysics explores questions such as how proteins in the eye respond to different energies of light, how electrical signals from the retina are transmitted to the brain, and how the eye muscles move the eye, focus the eye, and adjust the amount of light entering the eye.

Sensory biophysics also includes research to develop artificial sensory organs, for example, hooking a camera up to the optic nerve or to the visual cortex of the brain so that a blind person can see. A cochlear implant is a similar device but for sound. The implant does not amplify sound, but rather detects sound and stimulates the auditory nerves in response to that sound.

Environmental Biophysics

Environmental biophysics focuses on the physical aspects of the relationship between the organisms and their environment. Of particular importance is the understanding of the natural flow of energy in the environment and factors that can affect that flow. The primary source of energy in the environment is solar energy, which is captured through the process of photosynthesis and converted by plants into food. There are other areas of focus within environmental biophysics as well.

Heat and Temperature Environmental Biophysics

This branch of environmental biophysics studies factors that control the availability of thermal energy (heat) within the environment. In this context, heat is measured as the temperature of the atmosphere, soil, and bodies of water (rivers, lakes, ponds, streams, and oceans). Organisms need thermal energy to function properly. Also, the availability and movement of heat within an environment has a strong influence on many processes happening in that environment. Those processes in turn affect the environment and the organisms within it. The effect of temperature on organisms and their environment is also studied within this branch of biophysics, including how organisms deal with excess heat, or lack of heat, and how organisms themselves affect local environmental temperatures (for example, the effects of leaf canopies on local forest temperature and the effects of microorganisms on soil temperature). Mathematical modeling is used to predict temperatures and their effects.

Resource and Mass Exchange Environmental Biophysics

Environment biophysicists use mathematical models to better understand the physical movement and availability of resources and materials within the environment. Examples of resources include heat (as just discussed), light, carbon, nitrogen, water, and air. Take, for example, carbon. Photosynthesis captures not only solar energy but carbon as well. Carbon is needed by all organisms. Plants capture carbon through photosynthesis, and animals get their carbon by eating plants or other animals.

Radiation Biophysics

Radiation biophysics studies the effects of radiation on biological systems. As a branch of biophysics, radiation biophysics can include the use of radiation in diagnostic imaging and treatment of diseases. Such uses of radiation typically fall under the category of imaging and medical biophysics. Within the context of environmental biophysics, however, radiation biophysics studies how organisms deal with radiation in their environments. Sources of radiation can be natural (for example, solar radiation) or human-made (for example, nuclear waste). Radiation can have a variety of effects on organisms and their environment. Solar radiation is the primary source of energy for living things, and also has an impact on the flow of heat, air, and water within the environment. Ultraviolet and other types of natural and human-made radiation can cause molecular damage. Most organisms have natural repair mechanisms that can often, but not always, repair damage to their cells. While unrepaired damage is often trivial, it can also result in positive mutations, as well as lethal tumors or adverse mutations.

Environmental Bioengineering

The engineering side of environmental biophysics includes

- Development of sensors to detect or measure various aspects of the environment
- Research and development of energy-saving materials and techniques
- Development of materials with little or no environmental impact
- Development of physical tools and techniques to counteract the negative environmental effects of various human activities
- Research and development into low-cost methods of recycling materials such as glass, metals, and plastics

Putting It All Together

Figure 2-6 summarizes the topics discussed in this chapter. You should study the figure to know the names of the topics and how they are categorized. Some topics fall into more than one division. For example, bioenergetics and biomechanics can both be studied at any level: subcellular, physiological, or environmental.

Many topics interact with each other and to some extent overlap. Here are some examples.

- Understanding conformational transitions requires a thorough understanding of molecular structure.
- The forces involved in ligand and molecular binding are the same types of forces that bind together different parts of a single molecule when the molecule folds up to take on a particular shape.

Summary of Biophysical Topics Categorized by Size of Subject

- Molecular and Subcellular Biophysics
    o The Structure and Conformation of Biological Molecules
    o Structure Function Relationships
    o Conformational Transitions
    o Ligand Binding and Intermolecular Binding
    o Diffusion and Molecular Transport
    o Membrane Biophysics
    o DNA and Nucleic Acid Biophysics
    o Protein Biophysics
    o Energy Flow and Bioenergetics
    o Thermodynamics
    o Statistical Mechanics
    o Kinetics
    o Molecular Machines
    o Allosterics

- Physiological and Anatomical Biophysics
    o Biomechanics
    o Electrophysiology
    o Sensory Biophysics

- Environmental Biophysics
    o Heat and Temperature Environmental Biophysics
    o Resource and Mass Exchange Environmental Biophysics
    o Radiation Biophysics
    o Environmental Bioengineering

FIGURE 2-6 • Summary of biophysical topics categorized by size of subject.

- Allosterics results from a combination of ligand binding and conformational transitions at the same time. And allosteric regulation is detectable by observing the kinetics of the process.

- Molecular transport often involves investigating how certain molecules get from one side of a membrane to another. And, the mechanisms of electrophysiology always involve molecular transport across membranes within excitable tissue.

QUIZ

Refer to the text in this chapter if necessary. Answers are in the back of the book.

1. What are the three major divisions of biophysics when categorizing the branches of biophysics according to size of what is being studied?
   A. Small, medium, and large biophysics
   B. Microscopic, tissue, and organism biophysics
   C. Subcellular, physiological, and environmental biophysics
   D. Subcellular, structural, and macromolecular biophysics

2. A conformational transition is
   A. measurable in a variety of ways.
   B. a change in the shape of a molecule.
   C. a way that biological molecules carry out their function.
   D. all of the above

3. Which of the following statements is *most* true?
   A. Secondary structure is two dimensional, whereas tertiary structure is three dimensional.
   B. Primary structure is one dimensional, specifying only the sequence in which atoms or groups of atoms are connected to one another.
   C. Biological molecules are rarely polymers.
   D. Residues are the parts of a molecule left behind after the molecule carries out its function.

4. A ligand is
   A. connective tissue that holds bones together.
   B. connective tissue that ties muscles to bones.
   C. an element in the periodic table.
   D. a smaller molecule or atom that binds to a larger molecule.

5. Name three branches of biophysics within the division of molecular and subcellular biophysics.
   A. Kinetics, membrane biophysics, and histology
   B. Sensory biophysics, radiation biophysics, and allosterics
   C. Allosterics, conformational transitions, and statistical thermodynamics
   D. Protein biophysics, orgone biophysics, and electrophysiology

6. Statistical mechanics falls into which division of biophysics?
   A. Molecular and subcellular biophysics
   B. Physiological biophysics
   C. Environmental biophysics
   D. All of the above

7. Cell membranes are primarily made of
   A. DNA.
   B. protein.
   C. polysaccharides.
   D. lipids.

8. Bioenergetics falls into which division of biophysics?
   A. Molecular and subcellular biophysics
   B. Physiological biophysics
   C. Environmental biophysics
   D. All of the above

9. Which of the following statements is *most* true?
   A. Statistical mechanics uses statistics to calculate the average force animals use to acceler-ate to their top speed.
   B. Statistical mechanics is the application of statistical methods to biomechanics.
   C. Statistical mechanics is a misnomer because in practice it has nothing to do with statistics.
   D. Statistical mechanics uses statistical averages of populations of molecules to calculate measurable thermodynamic quantities.

10. A biopolymer is a
    A. type of allosteric interaction.
    B. large collection of biomolecules within the cell.
    C. biological molecule made up of many smaller molecules linked together.
    D. lipid bilayer

chapter 3

Biophysical Techniques and Applications

In this chapter we present an overview of the various techniques and applications of biophysics. These are important primarily for two reasons. First, we need to know what we can learn from each technique and how each technique can be useful in studying the various topics of biophysics. Second, some of the techniques are branches of biophysics unto themselves.

CHAPTER OBJECTIVES

In this chapter, you will

- Learn to define the most common biophysical techniques.
- Understand what information each biophysical technique can provide.
- Be able to list and briefly describe several applications of biophysics in medicine and in other areas.

As we mentioned previously, biophysical techniques and applications are sometimes considered as branches of biophysics. That is, some biophysicists dedicate their focus of study to a single technique or application, with the following goals in mind: (1) to research and better understand the physical principles underlying the technique, and (2) to work toward extending the technique's abilities.

Biophysical techniques fall into two major categories: preparative and analytical. *Preparative techniques* are those that purify or isolate biological specimens (organisms, cells, and molecules) or otherwise get them ready for use in some other process or further experimentation. *Analytical techniques* are those used to measure physical aspects of a biological system. Many biophysical techniques fall into both categories at the same time.

Ultracentrifugation

A *centrifuge* is a machine used to spin a sample of material around in circles. The circular motion places a force on the sample. The force is similar to, but typically much larger than, the normal force of gravity. An *ultracentrifuge* is a centrifuge specially designed to spin at an extremely high rate of speed. Some ultracentrifuges can exert forces as much as 1 million times that of gravity.

Why do we do this? Centrifuges operate on the principle of sedimentation. *Sedimentation* describes the motion of particles in a fluid under the application of a force. Take for example a snow globe; you know, those glass or clear plastic containers filled with water and some sparkly snowlike particles. Typically the globe also contains some kind of winter scene. You shake up the particles inside and then put the snow globe down. The force of gravity causes the particles to slowly descend in the water, making it appear as though it is snowing inside the globe. This is sedimentation. In fact, real snow showers result from sedimentation of snowflakes in the atmosphere.

The physics of sedimentation shows that the sedimentation rate of a particle or molecule depends on several things, including the force, the density of the fluid, and the size and density (or concentration) of the particles in the fluid. Applying a force stronger than gravity can increase the sedimentation rate. It can also magnify differences in sedimentation behavior between different molecules. This makes ultracentrifugation a convenient technique for separating molecules of different sizes. Ultracentrifugation is used as both a preparative and an analytical technique. For example, the ultracentrifuge is commonly used to isolate samples of pure DNA.

An *analytical ultracentrifuge* contains special optical devices and sensors that can track the movement of molecules as they are being centrifuged. Sedimentation rates can be measured directly, under various conditions. We can use the formulas that describe the physics of sedimentation to calculate the size and approximate shape of the molecules. An analytical ultracentrifuge can also be used to detect conformational transitions (see Chap. 2) and to determine the number of subunits making up a molecular complex.

Electrophoresis

Electrophoresis is another technique that relies on the principle of sedimentation. However, in electrophoresis the force results from an electric field applied to electrically charged particles or molecules. Many biomolecules such as DNA and proteins have an electric charge. We can take advantage of this by applying a strong electric field to the molecules in solution. Under these conditions, for example, negatively charged DNA will move (sediment) through a solution toward the positive side of the applied electric field.

A very common type of electrophoresis is *gel electrophoresis*. A gel is a fluid that has a molecular structure that gives it properties similar to a solid. Jellies and jams are gels. The molecular structure that gives the gel its solid-like properties also acts to obstruct the movement of molecules dissolved in the gel. Larger molecules are more easily obstructed than smaller molecules, so the gel actually increases the differences in sedimentation rates between molecules of different sizes within the gel.

Think of balls rolling down a hill and imagine there are many obstacles on the hill (fence posts, trees, bushes, boxes, etc., anything that would obstruct the balls freely rolling down the hill). The balls bounce off the obstacles but eventually find their way to the spaces between the obstacles and continue rolling down the hill. But if the balls are of various sizes, then the larger balls—being larger and taking up more space—are more likely to bang into the obstacles along the way. This is especially true if the gaps between obstacles are not much bigger than the largest balls. Every time a ball hits an obstacle, it slows its descent.

The smaller balls, however, sail through the gaps between the obstacles and move to the bottom of the hill faster. In the same way smaller, more compact molecules move through a gel much faster than larger molecules.

Sedimentation in gel electrophoresis is affected not only by the density of the gel and the size and shape of the molecules but also by the charge on the molecules. Molecules with more charge will experience a stronger force generated by the electric field.

Size Exclusion Chromatography

Size exclusion chromatography (SEC) is another technique that relies on sedimentation. SEC, however, uses gravity, or sometimes pressure, to sediment a solution through a gel. The gels used in SEC, however, differ from those used in electrophoresis in the following way. SEC gels are not one solid piece of gel, but rather a tightly packed suspension of gel beads or particles with spaces between them. There are pores on the surface of the gel particles through which molecules pass to get into the gel matrix on the inside each particle or bead. These pores, however, are very small, excluding larger molecules. Smaller molecules enter these pores, but when they do it can take them some time to pass through or otherwise exit the gel bead. This slows down the smaller molecules. Larger molecules bump into the gel particles (as we saw with electrophoresis) but move around the spaces between the gel beads much faster than the smaller molecules that are temporarily trapped inside the gel beads. The end result is similar to electrophoresis in that molecules are separated based on their size. However, in SEC it is the larger molecules that pass through the gel faster and the smaller molecules that lag behind. Using gels with different size pores can "exclude" different size molecules from the insides of the gel particles.

Spectroscopy

As a general class of investigative techniques, *spectroscopy* typically involves sending some form of electromagnetic radiation into a sample and measuring various properties of the electromagnetic radiation that emerges from the sample. For example, one can measure the intensity of the emergent radiation. Other common properties include the direction of the emitted radiation and its polarization. The measured property is then plotted as a function of the wavelength or frequency of the radiation; the resulting plot is called a *spectrum* (plural: *spectra*).

Just as a reminder, all electromagnetic radiation can be characterized by wavelength or frequency. The two are inversely proportional. If we know one, we can calculate the other.

$$\upsilon = c/\lambda \tag{3-1}$$

where υ is the frequency, c is the speed of light (2.9979×10^8 m/s), and λ is the wavelength. So, plotting a spectrum by wavelength or frequency is more or less the same thing. It certainly contains the same information. The frequency, however, has the advantage that it is proportional to the energy; thus

$$E = h\upsilon \tag{3-2}$$

where h is Planck's constant (6.626068×10^{-34} m²kg/s).

Originally, the techniques of spectroscopy were developed using only the visible light portion of the electromagnetic spectrum [wavelengths of approximately 380 to 750 nanometer (nm)], and later grew to include ultraviolet, infrared, and a much broader range of wavelengths.

Over the years, additional techniques have been developed which do not necessarily involve electromagnetic radiation but which nonetheless produce a spectrum of sorts. For example, electron spectroscopy measures the kinetic energy of electrons that emerge from the sample. Mass spectrometry produces a spectrum as a function of mass. By and large, however, most spectroscopic techniques involve electromagnetic radiation. Therefore, when necessary in order to distinguish those forms of spectroscopy that use electromagnetic radiation from other uses of the word *spectroscopy*, we will specifically use the term *EM spectroscopy*.

There are dozens of spectroscopic techniques used in the biophysical sciences. In the following sections we briefly describe a few of them. Each type of spectroscopy teaches us something different about the biological sample, with some overlap in what we can learn from each technique. Commonly, spectroscopic techniques provide information about the identity of biological molecules, about their structure, conformational transitions, binding, and kinetics.

The various spectroscopic techniques are classified according to the type of light (electromagnetic radiation) used and according to the properties of the emergent light measured. Additionally, some spectroscopic techniques are further distinguished according to the conditions of the experiment that are controlled. For example, if we measure the amount of light absorbed as we control and slowly vary the temperature of a sample, we would call this *temperature-scanning absorbance spectroscopy*.

A note on the use of the word *light*: Throughout this book we will often use the word *light* rather freely to mean electromagnetic radiation. Strictly speaking, the word *light* is meant to distinguish those parts of the electromagnetic spectrum that are visible to living organisms. In practice, light is something we are very familiar with, and use of the word *light* to mean electromagnetic radiation emphasizes the fact that all forms of EM radiation have the same basic nature that light has. That is, they are rapidly oscillating electric and magnetic fields of various frequencies (which we perceive as color) and amplitutdes (which we perceive as intensity or brightness). Therefore you can assume within context that *light* is meant to convey in general any portion or all of the electromagnetic spectrum, from gamma rays and X-rays, through visible light, microwaves, and even radio waves. Of course, wherever it is important to distinguish one portion of the electromagnetic spectrum from another or to make very clear which type of light we are talking about, then we will use specific qualifiers such as *infrared, visible, ultraviolet, X-rays,* and so on.

Absorption Spectroscopy

Absorption spectroscopy (also called *absorbance spectroscopy*) is one of the more common (and easy to understand) forms of spectroscopy. In *absorption spectroscopy,* we shine light of a specific wavelength through a sample and measure the intensity of the light that comes out the other side. Really we are most interested in measuring the *absorbance,* or how much light does *not* come through to the other side (i.e., how much light is absorbed).

The absorbance of a given sample depends on three things: (1) the intrinsic ability of the molecules in solution to absorb light, (2) the concentration of the molecules in solution, and (3) the path length of the light as it passes through the sample (i.e., if the sample container is larger, then the light has to pass through more of the solution before getting to the other side, and so more light will be absorbed). Since our goal is to know a given molecule's ability to absorb light, we need to somehow account for the concentration of the solution and the path length of the light. The *molar extinction coefficient* is a measure that accounts for both concentration and thickness of the sample being studied. It does this by expressing the absorbance in units that are per concentration and per length ($M^{-1}cm^{-1}$).

It is common to measure absorbance at many different wavelengths and plot a graph of the absorbance versus the wavelength or frequency of light. Such a graph is called the *absorption spectrum* (Fig. 3-1). Many molecules have unique or characteristic absorption spectra, so an absorption spectrum can be used to identify types of molecules in a sample. Absorption spectroscopy can also be used to measure the concentration of molecules in solution (once the identity of the molecule is known).

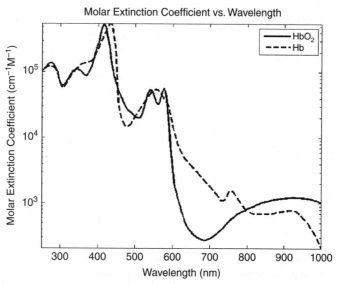

FIGURE 3-1 • Absorption spectra of hemoglobin with oxygen bound to the hemoglobin (solid line) and without oxygen bound to the hemoglobin (dashed line). (*Courtesy of Wikimedia Commons.*)

Notice in Fig. 3-1 that the difference in absorbance between the two forms of hemoglobin is greatest at a wavelength of about 690 nm. This means we can measure absorbance at 690 nm as a convenient way to determine the proportion of hemoglobin with oxygen bound to it. For example, let's say we have a sample of hemoglobin. Now say we measure the difference in absorbance between our sample and oxygen-free hemoglobin to be only 3/4 of the difference we expect for fully oxygenated hemoglobin. Then we can say that 75% of the oxygen binding sites in our hemoglobin sample have oxygen bound to them, while 25% remain unoccupied.

In many molecules, absorption of light also varies with conformation, as well as with the presence or absence of bound ligands. So absorption spectroscopy can be used to follow conformational transitions and ligand binding. Temperature-scanning absorption spectroscopy measures absorbance (usually at a single wavelength) across a range of temperatures. This is useful for studying temperature-induced conformational transitions, that is, changes in molecular shape that can be brought about by changes in temperature. This is a common technique for studying conformational transitions in membranes, proteins, and DNA (nucleic acids). See Fig. 3-2.

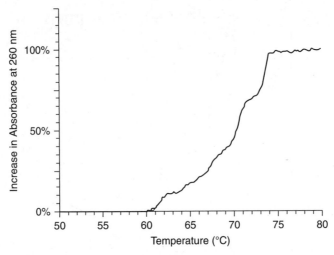

FIGURE 3-2 · A single-stranded DNA helix absorbs more light at 260 nm than a double helix. This makes it possible to use absorption at 260 nm to measure unwinding of the DNA double helix into individual single strands of DNA. The conformational transition from double helix to single-stranded DNA can be induced by increasing temperature. This makes temperature-scanning absorption spectroscopy a convenient tool for studying these conformational transitions in DNA.

Fluorescence Spectroscopy

Fluorescence is the opposite of absorption. In absorption, light is converted into the kinetic energy of electrons in an atom or molecule; this bumps the electrons into a higher or excited energy state. In fluorescence, the electrons drop down from their excited state, emitting light in the process. Fluorescence is caused by absorption, although not all absorption results in fluorescence. The wavelength of the emitted light is typically longer than that of the absorbed light.

Fluorescence spectroscopy, as with absorption spectroscopy, can be used to characterize (i.e., identify) molecules and to measure and follow conformational transitions and ligand binding. A *fluorophore* is a small molecule or the specific part of a molecule that is responsible for the fluorescence. *Fluorescent tagging* is a technique in which a fluorophore is attached to another molecule in order to track that molecule through some biological process. Fluorescent tagging is one of the techniques used to determine the sequence of residues in DNA.

Mass Spectrometry

Mass spectrometry is a technique in which molecules or parts of molecules are ionized and then passed through a magnetic field. By measuring the movement of the charged molecules in the magnetic field, it is possible to get a very accurate measurement of the mass or molecular weight of the molecules.

Mass spectrometry is used both for determining molecular weights and for identifying molecules (once the molecular weight is known). When combined with other methods, mass spectrometry can also reveal information about the structure of molecules. The very large molecules typically found in biological systems present particular challenges. Large molecules can be difficult to ionize in a quantifiable way, and it can be difficult to cause them to fly through a vacuum (a requirement for mass spectrometry). However, recent breakthroughs are making the use of mass spectrometry in biophysics quite fruitful.

X-Ray Crystallography

X-ray crystallography is a technique for determining the relative positions of atoms within a crystal. A *crystal* is an orderly, three-dimensional, repeating arrangement of atoms or molecules. Many substances can be crystallized; this means that conditions can be arranged so the intermolecular forces cause the molecules to line up in an organized, repeating manner.

The technique of X-ray crystallography provides very precise, high-resolution structural information for molecules in a crystal. For example, X-ray crystallography was used to discover that DNA is a double helix. The great advantage of X-ray crystallography is the high resolution of structural detail that it can provide (see Fig. 3-3). The disadvantage is that the molecules must be in crystalline form in order for the technique to work. Fortunately, over the years, crystallographers have become quite adept at manipulating chemical conditions in order to cause many biomolecules to crystallize. Still, not all molecules can be crystallized, in which case other techniques must be used to obtain structural information.

X-ray crystallography operates on the principle of diffraction. *Diffraction* occurs when light waves pass through an ordered arrangement of openings (as one finds in a crystal) and interfere with each other on the other side. Each opening acts as a new starting point for the waves. The waves from each opening then meet on the other side. When waves meet in phase with one another, the peaks from one wave come together with peaks from the other wave, and troughs from one wave come together with the troughs from the other wave. The result is *constructive interference;*

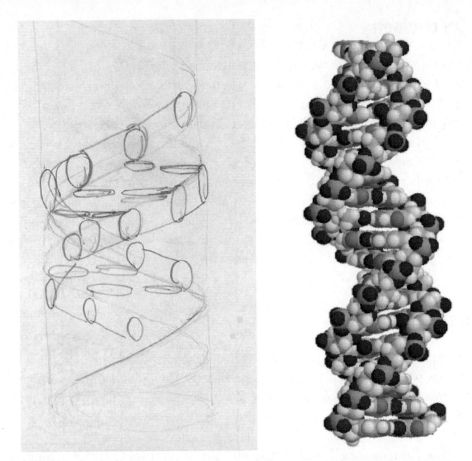

FIGURE 3-3 • DNA double helix. On the left is the original sketch by Francis Crick on discovering the DNA double helix in 1953. On the right is a space-filling model showing the various atoms. Both were determined using X-ray crystallographic data.

the peaks become higher and the troughs become lower. But when waves meet out of phase, a peak and a trough come together. The result is *destructive interference;* the waves flatten out and lower the intensity of radiation.

The technique takes advantage of the behavior of electromagnetic waves when they encounter atoms or molecules that are arranged in a regular, repeating structure. The atoms in general change the direction of the electromagnetic wave fronts, scattering them in all directions. The result is that each atom in the crystal acts as a starting point for the scattered waves. The scattered waves then interfere with each other, resulting in a pattern of constructive and destructive interference at various locations.

The regular repeating pattern of atoms simplifies that mathematics. If we measure the angle of the incoming radiation, the wavelength of the radiation, and the distances and angles to the points where we observe constructive interference, we can then extrapolate back to the source of scattering (the atoms) and calculate the distances between the atoms.

X-rays, as you know, are a type of electromagnetic radiation, as is light. But X-rays are not visible to the eye. Instead we use photographic film and other detection devices in order to "see" X-rays and measure their intensity.

Still Struggling?

Mathematically, electromagnetic radiation can be treated as particles, called *photons,* or waves (electromagnetic waves). This is sometimes referred to as the dual nature of light: light appears to behave as particles in some situations and as waves in others. We use the particle model or the wave model, depending on which best explains the phenomenon we are discussing. In some cases either will do. But in other cases one is clearly more convenient than the other. As particles, photons exhibit elastic collisions, bouncing, and scattering off of objects. (They can also be absorbed. See absorption spectroscopy.) As waves, however, electromagnetic radiation exhibits *reflection* (bouncing off a surface), *refraction* (bending while passing through a surface), and *interference* (the combining of two or more waves in a specific location, resulting in increased or decreased wave height at that location).

Nuclear Magnetic Resonance Spectroscopy

Nuclear magnetic resonance (NMR) spectroscopy is a type of EM spectroscopy that differs from most forms of EM spectroscopy in the following significant and practical ways:

- NMR involves interaction of light (EM) with the nuclei of atoms in a molecule, whereas most forms of EM spectroscopy involve interaction of light with the electrons in the molecule.
- Although most forms of spectroscopy do provide some structural information, NMR can provide much more structural details (higher resolution) than other forms of spectroscopy.
- NMR involves the application of a strong magnetic field to the sample being studied. This magnetic field alters and limits some of the energy

states available to the nuclei; this is what makes it possible to measure the absorption and emission of EM by the nuclei in the sample.

- NMR uses EM in the radio frequency (RF) portion of the spectrum (whereas, for example, absorption and fluorescent spectroscopy typically rely on EM in ultraviolet, visible, and infrared regions of the EM spectrum).

NMR works on the basic principle that a spinning charge (such as the nucleus of an atom) generates a magnetic field. In other words the nucleus is like a small magnet. Under normal conditions, the spins of the various nuclei are randomly oriented in any direction. However, under the influence of a strong magnetic field the nuclear spins are constrained to only certain orientations with respect to the external magnetic field (typically parallel and antiparallel to the magnetic field).

When a nucleus jumps from one spin orientation to another, it will absorb or emit EM radiation. The frequency of this EM radiation will be proportional to the energy difference between the two spin states. By scanning the EM spectrum, we can find all the specific frequencies at which the nuclei are absorbing and emitting radiation, and thus determine all the energy differences between spin states. Each of these energy differences depends on the strength of the magnetic field in the local region of the molecule surrounding the nucleus. Nuclei that are shielded by electrons and other atoms will experience less of the applied magnetic field, so the energy difference between their spin states will be smaller. The smaller energy difference between spin states means that these nuclei will resonate with lower frequencies (less energy) of EM. On the other hand, the nuclei that are less shielded by electrons and other atoms will be more exposed to the magnetic field. These nuclei resonate with higher-frequency EM. Therefore, by observing which frequencies of light are absorbed and emitted as a result of the magnetic field, we can infer structural information about molecules.

Electron Microscopy

The most powerful light microscopes provide only enough magnification to view objects larger than 200 nm. This limitation is due to the wavelength of visible light. In 1928 physicist Ernst Ruska was experimenting with magnetic lenses for focusing electron beams and realized that it was possible to take advantage of the smaller wavelength of electrons to create an imaging device theoretically capable of greater magnification than a light microscope. In 1931 Ruska and fellow engineer Max Knoll built the first electron microscope. Although it was no more powerful than a light microscope, they had proved the concept of using focused electron beams for microscopic imaging. By 1933 they built an electron microscope that surpassed the resolving power of light microscopes. Today electron microscopes are capable of viewing objects 1000 to 2500 smaller than what can be seen with even the most powerful light microscope.

Whereas a light microscope focuses a beam of light on the specimen to be magnified, electron microscopes use a beam of highly accelerated electrons. There are several types of electron microscopy. The two most common types are transmission electron microscopy and scanning electron microscopy. *Transmission electron microscopy (TEM)* is similar to light microscopy in that the beam passes through a very thinly sliced sample to provide an image on the other side. Since electrons are not visible to the eye, the image is made by focusing the electron beam onto a view screen coated with some material that fluoresces (emits visible light) in response to the incoming electrons. The image can be further enhanced using detectors similar to those found in digital cameras.

In *scanning electron microscopy (SEM)* the electron beam is gradually scanned across the surface of the specimen. At each point on the surface some of the intensity of the electron beam is lost. This decrease of intensity can be measured in a variety of ways and can be converted into a surface image of the object (see Fig. 3-4). SEM is only about 1/10 as powerful as TEM, but it provides a more three-dimensional image instead of a cross-sectional slice of the specimen.

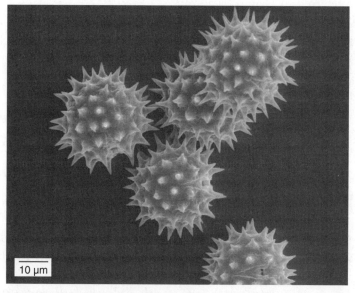

FIGURE 3-4 · Scanning electron micrograph of *helianthus annuus* (sunflower) pollen. (*Courtesy of Louisa Howard*)

Atomic Force Microscopy

Atomic force microscopy (AFM) provides magnified images similar to those of SEM, but with resolution similar to those of TEM (i.e., about 10 times better than SEM). AFM works by moving a mechanical probe across the surface of the object being scanned. AFM therefore provides true, three-dimensional information about the object. There are several variations of AFM that vary somewhat in how data are collected and in how the probe is manipulated. All are part of the more general class of techniques called *scanning probe microscopy (SPM)*. The more advanced techniques of SPM and AFM keep the probe a constant distance from the object's surface in order to avoid the possibility of the probe damaging or deforming the surface of the object being scanned.

Optical Tweezers

An *optical tweezers* is an instrument that uses focused laser beams to create piconewton (10^{-12} N)-size forces that can be used to hold and manipulate microscopic particles, even as small as a single molecule or atom. The phenomenon of focused laser beams holding a single particle in place in three dimensions is called an *optical trap*. For this reason, an optical tweezers is also sometimes referred to as an optical trap.

Optical tweezers can be used to hold and manipulate particles anywhere from 0.1 nm (about the size of an atom) to 10,000 nm in size (about the size of a bacterium), and have been used to trap single viruses, DNA molecules, bacteria, living cells, and organelles. Optical traps are particularly useful for investigating the mechanics of and forces associated with molecular motors. They can be used for sorting and separating cells of various types or for measuring the forces necessary for bending or breaking a DNA molecule.

Voltage Clamp

A *voltage clamp* is a technique used in electrophysiology to measure and characterize electric currents in cells, particularly in excitable tissue cells such as neurons (nerve cells). The basis of the technique is the ability to insert a very fine microelectrode into the cell, with another electrode in contact with the fluid around the outside of the cell. The electrodes can be used to measure (and manipulate) the voltage or current across the cell membrane. In the voltage clamp technique, the voltage is clamped, or held constant, through a feedback mechanism. The value of the voltage that is held constant is called the *command voltage*. The feedback mechanism detects even the slightest change in the voltage and immediately pumps current across the membrane through the electrode to keep the voltage at the command voltage. To do this, the current generated by the electrode needs to be equal and opposite to the current generated by the cell. We record the current generated by the feedback mechanism and use it to calculate the current generated by the cell.

A voltage clamp allows us to measure the cell's electric current under a variety of conditions. If we do this for a range of command voltages, then we can also measure how voltage affects the current. This is useful to discover and characterize *voltage-gated ion channels* which are proteins found in some cell membranes that allow ions to pass through the membrane only when the voltage is within a particular range.

Current Clamp

The *current clamp* is a technique similar to the voltage clamp, except that the electrode feedback mechanism is used to keep the current (across the membrane) constant while allowing the voltage to vary. In this way we can measure how the cell varies the voltage across its membrane in response to a particular current or other conditions.

Patch Clamp

The *patch clamp* is an alternative technique for applying the electrode to the cell. Instead of poking a sharp, metal electrode through the cell membrane, the electrode is placed inside a micropipette (a very thin glass tube) filled with an electrolyte solution, and the micropipette is placed against the cell membrane. Gentle suction is applied, forming a tight seal between the micropipette and the cell membrane (see Fig. 3-5). This tight seal has a high electrical resistance [1 to 10 megaohms ($M\Omega$)] electrically isolating the small patch of membrane under the electrode. This allows us to investigate the behavior of a single ion channel within the membrane.

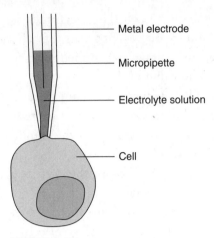

FIGURE 3-5 • Patch clamp technique: Instead of poking a sharp metal electrode through the cell membrane, the electrode is placed into a micropipette (a thin glass tube) and the micropipette is held against the cell membrane.

Still Struggling

The terms voltage clamp, current clamp, and patch clamp all sound like devices rather than techniques. The device, however, is the same for all three. The device consists of an electrode, some wires, and an electronic feedback unit that is capable of simultaneously measuring and manipulating voltages and currents. The techniques differ primarily in the settings on the device: whether to measure current or voltage, which to hold fixed (and at what value), and which to manipulate via the feedback mechanism. In the case of the patch clamp, there is the addition of a micropipette (a thin glass tube, commonly found in any physiology laboratory) into which the electrode is placed.

Calorimetry

Calorimetry is the measurement of energy changes in the form of heat. An instrument designed to measure heat energy is called a *calorimeter*. In biophysics, calorimetry is typically used to measure the amount of energy absorbed or released in the course of biochemical reactions, conformational transitions, or ligand binding. For example, calorimetry can be used to determine the amount of energy necessary to unwind a piece of DNA helix. Calorimetry can also be used to measure the binding strength of various drugs to a particular protein. The results are then used to determine which drug is most effective at a particular task. There are various types of calorimeters. In particular, microcalorimeters are useful for measuring the very small amounts of energy associated with many biophysical processes.

There are numerous other biophysical techniques not yet discussed here. The goal of this chapter has been to present and define some of the more common techniques.

QUIZ

Refer to the text in this chapter if necessary. Answers are in the back of the book.

1. An ultracentrifuge operates on the principle of
   A. electrophoresis.
   B. spinning very fast.
   C. sedimentation.
   D. gravity.

2. The various forms of EM spectroscopy typically measure some property of the EM radiation as a function of
   A. wavelength.
   B. intensity.
   C. speed of light.
   D. density.

3. Oxygen-free hemoglobin absorbs 10 times more light at 690 nm than fully oxygenated hemoglobin. If a sample of hemoglobin is found to absorb only 5 times as much light as fully oxygenated hemoglobin, what percentage of the hemoglobin molecules have oxygen bound to them? (Assume that each hemoglobin molecule is either fully bound or fully unbound.)
   A. 5%
   B. 10%
   C. 50%
   D. 690%

4. Fluorescence is the opposite of
   A. darkness.
   B. incandescence.
   C. absorption.
   D. X-ray crystallography.

5. A mass spectrograph characterizes molecules according to their molecular weight by
   A. placing them on a scale.
   B. weighing them in a vacuum.
   C. ionizing them and passing them through a magnetic field in a vacuum.
   D. ionizing them and passing them through an electric field in a gel.

6. In order for X-ray crystallography to work,
   A. the X-rays must be focused using a glass crystal.
   B. the X-rays must be bound to the molecules being studied.
   C. the molecules must be in a regular, repeating arrangement.
   D. the molecules must not cause electromagnetic interference.

7. Which of the following statements is *most* true?

   A. NMR is preferred over X-ray crystallography because the resolution is higher with NMR.
   B. X-ray crystallography provides very precise, high-resolution, structural information.
   C. Optical tweezers can be used to place a voltage clamp on the optic nerve.
   D. AFM works in theory, but has yet to be proved.

8. Gel electrophoresis

   A. operates on the principle of sedimentation.
   B. can be used to separate molecules on the basis of size.
   C. is both an analytical and a preparative technique.
   D. all of the above

9. The sedimentation force in gel electrophoresis comes from

   A. the sedimentation coefficient.
   B. gravity.
   C. an electric field.
   D. all of the above

10. Which of the following statements is *most* true?

    A. A calorimeter is a convenient way to count calories.
    B. A microcalorimeter can measure very small amounts of heat.
    C. A limitation of calorimetry is that it cannot be used to measure the heat of conformational transitions.
    D. A microcalorimeter can fit on the head of a pin.

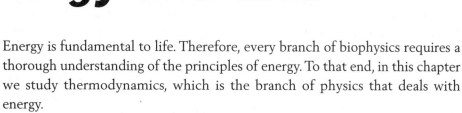

4

Energy and Life

Energy is fundamental to life. Therefore, every branch of biophysics requires a thorough understanding of the principles of energy. To that end, in this chapter we study thermodynamics, which is the branch of physics that deals with energy.

CHAPTER OBJECTIVES

In this chapter, you will

- Learn the laws of thermodynamics as they apply to biophysical systems.
- Be able to explain how Gibbs energy determines whether a biophysical process will take place.

The First Law of Thermodynamics

Thermodynamics is the study of energy and how it operates in the physical universe. Statistical mechanics (which we review in the next chapter) is closely related to thermodynamics. *Statistical mechanics* is the study of how probabilities and statistical averages of particles can be related to the overall thermodynamic measurements of a system. For example, the average speed (or average kinetic energy) of water molecules in a glass of water is directly related to the temperature of the water. In biophysics, the particles whose statistics we are interested in are biomolecules or, sometimes, residues and subunits within biomolecules.

The first law of thermodynamics is simple and intuitive. It states that any change to the amount of energy contained in a system is equal to the amount of energy put into the system minus the amount of energy taken out of the system.

$$\Delta E_{sys} = E_{in} - E_{out} \tag{4-1}$$

This makes sense and is just another way of saying that energy cannot be created or destroyed. As an analogy, imagine a bucket of marbles. The first law says that any change to the number of marbles in the bucket is equal to the number of marbles we put into the bucket minus the number of marbles we take out.

What is less intuitively obvious is the more conventional way to write Eq. (4-1). In practice we don't usually break up the right side of the equation in terms of whether the energy is coming in or going out. Instead we break it up in terms of the type of energy entering or leaving our system and let the sign on each energy term (positive or negative) indicate the direction of energy flow. By convention we write the first law of thermodynamics like this.

$$\Delta U = q - w \tag{4-2}$$

where U is the symbol for the *internal energy* of a system (what we previously called E_{sys}), q represents heat put into the system, and w is work done by the system.

Equation (4-2) presents the first law in terms of *heat energy* on the one hand, and *all other forms* of energy on the other, which we lump together into a single term called *work*. This is like saying that any change in the number of marbles in our bucket is equal to specifically the number of yellow marbles we put into the bucket minus the number of all other marbles we take out of our bucket. This is not the most obvious way to break things up. At least, not until you consider that maybe there's something special about yellow marbles.

Indeed this is the case: there is something special about heat as a form of energy. In fact there are three reasons why heat is special. It is primarily the first of these three reasons that explains why, by convention, we account for heat separately from other forms of energy. But the other reasons also provide motivation for thinking of heat separately.

Heat Is Special

The first reason we account for heat separately is historical. The laws and equations of thermodynamics were originally developed through the study of heat engines. A *heat engine* is a motor that converts heat energy into motion, for example, a steam engine or an internal combustion engine. This is why the conventional way to write the first law of thermodynamics is in terms of heat put into a system and work (motion) done by the system. Historically, almost all of our early experiences with thermodynamics involved lighting fires, boiling water, and trying to get some useful work out if it.

The second reason heat is special is that heat is the only type of energy that is always readily available to measure and possibly use. This might not seem the case on a very cold winter's day, but it's true. *Heat* is the kinetic energy and random motion of molecules. Even on a cold winter's day molecules are constantly vibrating or moving about. Even in solid ice, the crystallized water molecules are continually vibrating. Molecular kinetic energy (heat), as we will see, can contribute to biochemical processes. Molecular motion is always present. Only at a temperature of absolute zero ($-273.15°C$)—far colder than the coldest of winter days (and certainly far colder than we ever expect to find any living things that we study in biophysics)—does molecular motion stop.

The third reason heat is special is that heat is, in a sense, the lowest form of energy. Heat is the least organized form of energy. It is this disorganized quality of heat that also makes it often the most inefficient form of energy from which to extract useful work.

What Is a System?

All of the above assumes we know what we mean by *system*. In thermodynamics, a *system* is simply that part of the universe that we are interested in. The rest of the universe is called the *surroundings*. Thermodynamic systems are further qualified in terms of their contact with the surroundings. An *open system* can exchange both matter and energy with its surroundings. A *closed system* can exchange energy, but not matter. An *adiabatic system* is thermally isolated from its surroundings; there is no exchange of heat or matter, but it is possible for the system to do work on its surroundings (or to have work done on it).

In thermodynamics energy is typically measured in units of calories or joules. One calorie is the amount of energy it takes to increase the temperature of 1 g of water from 14.5°C to 15.5°C at 1 atmosphere of pressure. One joule is the amount of energy it takes to lift an object weighing 1 N a distance of 1 m. That is, 1 joule (J) = 1 newton meter (Nm). One calorie is equal to 4.184 J.

PROBLEM 4-1

A 225-lb athlete drinks a glass of orange juice (120 calories) and walks up to the top of the Empire State Building. What is the change in his internal energy, assuming the only heat transfer is the 120 calories from the orange juice, and the only work done by the athlete is lifting his own weight to the 102nd floor? Assume 3 m per floor.

SOLUTION

We are going to use Eq. (4-2) to solve the problem. To begin with, however, we need to know that food calories are actually kilocalories. So when the problem says a glass of orange juice has 120 calories, then in terms of the scientific unit of energy called *calorie,* this is 120,000 calories (cal), or 120 kilocalories (kcal).

The amount of heat entering our athlete is 120 kcal, or 502 kJ (120 kcal × 4.184 kJ/kcal). Work is force times distance. Therefore the amount of work done is the weight of the athlete (the force) times the distance he is lifted. Converting to SI units, 225 lb is about 1000 N of force. So the amount of work to lift the athlete to the 102nd floor is

1000 N × (102 floors × 3 m/floor) = 306,000 Nm = 306 kJ of work.

The change in the athlete's internal energy is

$$\Delta U = q - w = 502 \text{ kJ} - 306 \text{ kJ} = 196 \text{ kJ}$$

The athlete's internal energy has increased (by 196 kJ) because the glass of orange juice contained more energy than is necessary to lift 225 lb to the

top of the Empire State Building. Our athlete used up less than 2/3 the calories in the glass of orange juice.

If you have some concept as to what it feels like to walk up almost 2000 steps, and how you might feel after such a workout, then it may seem odd that walking up 102 flights of stairs burns off fewer calories than are in 2/3 of a glass of orange juice. The apparent discrepancy (from our everyday experience) here is that we have made some simplifying assumptions that perhaps include a little too much simplification.

Simplifying Assumptions

In science, we often make simplifying assumptions. We purposely assume something to be the case, even though we know it's not true, in order to keep things simple. Keeping things simple can allow us to make a calculation or discover something that we would otherwise not be able to do or discover.

Then, after we understand our system better, after we have calculated what we need to calculate, we can (if we choose) refine our calculations and refine our understanding by gradually removing or changing our simplifying assumptions. Of course, the most important aspect of this is to know what our assumptions are and to make a judgment as to whether the assumptions are reasonable or at least practically workable. By practically workable we mean that (1) even if our assumptions are wrong, we understand that they are wrong and we have some idea where they are off (or where they differ from the more correct solution) and (2) most importantly the simplifying assumptions allow us to accomplish something that we would otherwise not be able to accomplish without their help.

As an example of accomplishing something practical even with assumptions that are incorrect, imagine if we were to measure the energy content in food by drinking a glass of orange juice, doing some work, and seeing how much work we can get done before we get hungry again. If the athlete in Prob. 4-1 got hungry when he reached the top of the Empire State Building, we would say that the orange juice had 306 kJ of energy. To say such a thing, we have to make some simplifying assumptions, not the least of which is the idea that the athlete could tell by his hunger when he has burned off the glass of orange juice. Even so, this method still gives us some clue as to how much energy is in the orange juice. Assuming the athlete can achieve some accuracy in reporting his hunger, even if the resulting energy content is way off, we can still compare various foods, and get an idea of their energy content relative to one another. Or, at the very least we have a measure of how well various foods (relative to one another) stave off hunger during exercise, which is somewhat related to energy content of the food.

This example is perhaps a bit far fetched, but you get the idea. Making simplifying assumptions is a useful tool. Later we will see more practical examples of how making simplifying assumptions enables us to accomplish many things.

So what are the simplifying assumptions that we made in Prob. 4-1? Here are a few of them. You should try to think of others. The first and probably largest assumption is that 100% of the energy of the orange juice is converted into work done by our athlete. As a general rule, animals are not the most efficient heat engines. Humans, at best, convert only about 10% of their caloric intake into useful work. The majority of the energy content of the food we eat is either dissipated as heat or excreted as waste.

Another assumption we made, explicitly mentioned in the problem, is that the only work done by the athlete is lifting his own weight to the 102nd floor. This is not entirely correct. There is also work done to pump blood around the body and work to expand and contract the lungs. And there is even some work overcoming friction of the athlete's shoes against the floor and of air resistance while walking up stairs.

What other assumptions did we make? See if you can come up with two or three additional assumptions, perhaps not as big and significant (in terms of the calculated result) as those just mentioned, but assumptions none-the-less. (See the Quiz at the end of this chapter for a few more assumptions.)

Enthalpy

Enthalpy is a classic thermodynamic property of a system, defined as the sum of the internal energy of a system plus its pressure times its volume.

$$H = U + PV \tag{4-3}$$

Although it may not be intuitively obvious how such a quantity can be useful, we will see later that enthalpy can be easy to measure and it plays a role in calculating whether a given biological process will occur spontaneously.

You can think of a system's enthalpy in much the same way you think of heat. A change in enthalpy ΔH is very similar to q, heat flowing into the system (or flowing out if q is negative). The main difference between heat and enthalpy is in the way we account for changes in pressure. If you want, you can skip over the remainder of this section and start reading the section on Entropy, without losing any understanding toward our chapter objectives. The rest of this section provides experience with manipulating thermodynamic quantities and can give you a deeper understanding into the nature of enthalpy.

By the definition of enthalpy, Eq. (4-3), the change in enthalpy is equal to the change in internal energy plus the change in the product of pressure and volume.

$$\Delta H = \Delta U + \Delta(PV) \tag{4-4}$$

We can see the relationship between enthalpy and heat by rearranging Eq. (4-4).

First, we rewrite it in a form similar to Eq. (4-2), in which we expressed the first law of thermodynamics with the internal energy on the left and all other terms on the right.

$$\Delta U = \Delta H - \Delta(PV) \tag{4-5}$$

Next, we account for pressure and volume changes in separate terms. That is, we can rewrite $\Delta(PV)$ as

$$\Delta(PV) = P\,\Delta V + V\,\Delta P \tag{4-6}$$

This step may seem a bit like hand waving to you or may be familiar (if you are familiar with calculus or differential equations). The general principle is that the change of a product or two or more variables, $\Delta(PV)$ in our example, is equal to a sum of two or more products, where each product contains the change in only one variable multiplied by the value of each of the other variables held constant. Thus $\Delta(ABC)$ for example, would be equal to $(\Delta A)BC + A(\Delta B)C + AB(\Delta C)$. We will not take the time to prove this mathematic concept here, but we do demonstrate its truthfulness by way of example in Probl. 4-2.

If we take the right side of Eq. (4-6) and substitute it for $\Delta(PV)$ in Eq. (4-5), we get

$$\Delta U = \Delta H - V\,\Delta P - P\,\Delta V \tag{4-7}$$

The last term on the right $P\,\Delta V$ is the work done by the system. We save the proof for Prob. 4-3. In short, since work is defined as force times distance, pressure times a change in volume is just the three-dimensional version of force times distance. Therefore we can rewrite Eq. (4-7) as

$$\Delta U = \Delta H - V\,\Delta P - w \tag{4-8}$$

Comparing this with Eq. (4-2),

$$\Delta U = q - w$$

we see that Eqs. (4-8) and (4-2) are equivalent if

$$q = \Delta H - V \Delta P$$

This gives us some insight into what enthalpy is; it's simply a way of dividing up heat into two different terms: one term is the energy associated with a change in pressure ($V \Delta P$) and the other term is the change in enthalpy (ΔH). Rearranging, we get

$$\Delta H = q + V \Delta P \qquad (4\text{-}9)$$

Enthalpy is just the heat flowing into the system plus the energy associated with a change in pressure.

Fortunately, almost all biological processes occur at constant pressure. At constant pressure, $\Delta P = 0$. This means that for a system at constant pressure, the enthalpy change is exactly equal to the heat flowing into the system: $q = \Delta H$.

Still Struggling

What is pressure times volume (PV)? How is $V \Delta P$ the energy associated with a change in pressure? Often we can understand a concept better by looking at the units of the quantities involved. The units of P are force per area (N/m²). If we multiply these units by meters per meter (which does not change the value at all because meters per meter is always equal to 1) we see that the units of P are Nm/m³. But a Nm is a joule, so the units of P are actually J/m³ or *energy per unit volume*. Pressure is energy per unit volume, a kind of energy density. When we multiple the pressure (J/m³) by the volume (m³), the units on PV are joules; PV is the potential energy associated with pressure, and $V \Delta P$ is the change in this energy due to a change in pressure.

PROBLEM 4-2

Choose example numbers to show that Eq. (4-6) gives correct results.

✔ **SOLUTION** _____

The equation is $\Delta(PV) = P\,\Delta V + V\,\Delta P$

The trick in solving this problem is knowing what value to use for the constant pressure or constant volume on the right side of the equation. Since both pressure and volume are changing, the appropriate value to use for the constant pressure and constant volume is the *average* value of the pressure or volume. Here we will make an assumption that the average value is equal to the midway point between the initial and final values. This assumption holds true under certain conditions, for example, if the pressure and volume change at a constant rate.

If we write each of the changes (denoted with Δ, the Greek letter delta) in terms of the difference between the final values and the initial values, we have

$$\Delta(PV) = P_2 V_2 - P_1 V_1$$
$$P\,\Delta V = P(V_2 - V_1)$$
$$V\,\Delta P = V(P_2 - P_1)$$

Now, to show Eq. (4-6) gives correct results, let's choose some example numbers. After reading this solution, you should try it yourself with your own numbers. Any numbers will do, but remember that, while the change in pressure or volume can be negative (meaning the pressure or volume decreased), each value of pressure or volume must be positive.

For our example, let's say the pressure went from 10 pascals (N/m^2) to 7 pascals, while the volume went from 13 m^3 to 19 m^3.

Substituting these numbers into the left side of Eq. (4-6), we get

$$\Delta(PV) = P_2 V_2 - P_1 V_1 = (7 \times 19) - (10 \times 13) = 133 - 130 = 3\ Nm = 3\ J$$

And substituting the same numbers into the right side, we get

$$P\,\Delta V = P_{avg}(V_2 - V_1) = [(10 + 7)/2](19 - 13) = 8.5(6) = 51\ J$$
$$V\,\Delta P = V_{avg}(P_2 - P_1) = [(13 + 19)/2](7 - 10) = 16(-3) = -48\ J$$

Then

$$P\,\Delta V + V\,\Delta P = 51\ J - 48\ J = 3\ J$$

which is the same as $\Delta(PV)$.

PROBLEM 4-3

Using the definition of pressure (force per unit area) and the definition of work (force times distance), show that $P \, \Delta V$ from Eq. (4-7) is just the three-dimensional work done by the system.

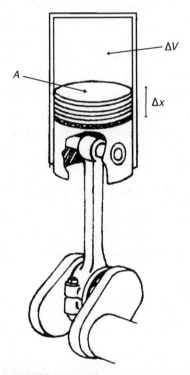

SOLUTION

Work, by definition, is force times distance moved:

$$w = F \, \Delta x$$

where Δx is the distance moved (change in position) along the same direction the force was applied.

Pressure, by definition, is force per unit area, or $P = F/A$.

To simplify the problem, consider a cylinder closed at one end, with a piston moving up or down inside the cylinder from the other end; see Fig. 4-1. The volume inside the cylinder (between the piston and the

FIGURE 4-1 · A piston in a cylinder allows us to simplify the case of work associated with pressure and a volume change.

closed end) changes as the piston moves up and down. The surface area of the piston is constant, so the volume change is just the area times the distance the piston is moved.

$$\Delta V = A\, \Delta x$$

So for the cylinder, we have

$$P\Delta V = \frac{F}{A}\Delta V = \frac{F}{A}(A\Delta x)$$

The *A* are canceled, leaving

$$P\, \Delta V = F\, \Delta x = w$$

In the case where work is done in multiple directions, we can imagine the work done in each direction as analogous to a piston in a cylinder aligned with that direction. Take, for example, the work done in compressing or expanding a balloon. We can imagine that at each infinitesimally small part of the balloon's surface there is a cylinder of constant area and the expansion or contraction of the balloon is equivalent to moving a piston up or down inside the cylinder. The total change in volume for the balloon is then the sum of all of these individual volume changes given by $\Delta V = A\, \Delta x$. Similarly the pressure against the surface of the balloon can be broken down, at each infinitesimally small part of the balloon's surface, as a force in a direction perpendicular to the balloon's surface at that point, divided by the area of the infinitesimal piston for which we are considering the volume change.

The volume change for any three-dimensional shape can, in this way, be broken down into a sum of many simple volume changes, each of which can be expressed as a constant area times a distance. And the pressure that causes that volume change is just, at each point on the surface, the force divided by that small constant area.

So we see that a constant pressure times a change in volume $P\, \Delta V$ is just the three-dimensional equivalent of force times distance, or work.

$$P\Delta V = \frac{F}{A}(A\Delta x) = F\, \Delta x \tag{4-10}$$

Entropy

The idea of entropy was developed by Rudolf Clausius in the mid-1800s. It was well known at the time that heat engines of the day converted only a small portion of their heat into useful work. Typically only about 1% of the heat was transformed. The rest of the heat was lost, dissipated, merely warming up the surroundings.

Clausius's goal was to account quantitatively for this loss. He originally called this lost energy the "equivalence value of all uncompensated transformations," meaning that it was equivalent to the portion of the original heat that could not be compensated for by transforming the resulting work back into heat.

A few years later he coined the term *entropy*. Clausius himself said he took the Greek word *trope* meaning "transformation" and put it in a form intended to resemble the word *energy*. So, from *trope* and *energy* he got *entropy*.

The change in entropy as the result of a process is defined as

$$\Delta S = Q/T \tag{4-11}$$

where S is the entropy and Q is the heat irreversibly lost to the surroundings at absolute temperature T. Notice that the units on entropy are energy units divided by temperature units, for example, joules per degree celsius.

If a process or transformation involves passing the heat through a working body (for example, a heat engine) so that the heat is passing from a place at temperature T_1 to a place at temperature T_2, then the entropy change is defined as

$$\Delta S = Q\left(\frac{1}{T_2} - \frac{1}{T_1}\right) \tag{4-12}$$

Although Clausius explained how he came up with the term *entropy*, he never indicated why he chose the letter S for entropy. Some have speculated that it was in honor of another famous early researcher of thermodynamics, Sadi Carnot. Figure 4-2 shows a diagram by Sadi Carnot illustrating the flow of heat and work in a heat engine.

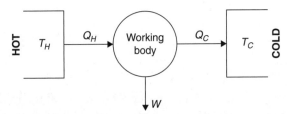

FIGURE 4-2 • Diagram of heat engine by Sadi Carnot from the year 1824.

PROBLEM **4-4**

A heat engine takes 400 J of heat and converts it into 3 J of work. Assuming that reversing the process will convert all of the work back into heat, how much of the original heat is not compensated for by reversing the process? What is the entropy change for this process at a temperature of 298 K?

✔ SOLUTION

The amount of heat not compensated for by reversing the process is

Uncompensated heat = original heat − work back into heat

The original heat is 400 J. Assuming reversing the process converts all the work back into heat, the heat compensated for by converting the work back into heat is 3 J. Therefore the uncompensated heat is

Uncompensated heat = 400 J − 3 J = 397 J

The entropy change, by definition, is

$$\Delta S = Q/T = 397 \text{ J}/298 \text{ K} = 1.33 \text{ J/K}$$

 PROBLEM **4-5**

If the heat engine from Prob. 4-4 transfers heat from a body at 350 K to a body at 300 K, what is the entropy change for this process?

✔ SOLUTION

From Prob. 4-5, the amount of heat lost is 397 J. The entropy change in going from 350 K to 300 K is then

$$\Delta S = (397 \text{ J})\left(\frac{1}{300 \text{ K}} - \frac{1}{350 \text{ K}}\right) = 0.189 \text{ J/K}$$

Entropy Is Related to the Number of Ways of Doing Something

It is worth mentioning here that entropy is related to order and disorder. You may have already heard this concept. Although Clausius originally defined entropy as a simple thermodynamic property (related to energy and work), research over the years has shown that entropy is actually a measure of *disorder* in a system.

But what do we mean by disorder? For now, let's just say that order and disorder in physics are *not*, as common sense might imply, a matter of neat versus messy. Rather, order and disorder relate directly to the *number of different ways of accomplishing one particular thing*. The more ways there are for a system to do one particular thing or to be in one particular state, the more disorder and the more entropy the system has.

For example, when there are only a few ways for the molecules of a system to arrange themselves to achieve a particular temperature, then we say that the system is highly *ordered*. But when there are many different ways for the same molecules to achieve a given temperature, then we say that the system is *disordered*.

Gibbs Energy

About 10 years after Clausius coined the term *entropy*, Josiah Willard Gibbs used Clausius's concept of entropy to define *available energy*, energy available to do useful work. Later Gibbs renamed it *free energy*, meaning the energy that is *free* to do useful work as compared with energy that is bound to be lost through dissipation as entropy. This latter way of thinking reflected the increasing realization that some heat energy is inevitably lost and simply can never be converted into work no matter how hard one tries. This idea that it is impossible to convert all of the energy into work without some loss gave rise to the second law of thermodynamics.

The free energy as Gibbs defined it is now called *Gibbs free energy*, or simply Gibbs energy, or the *Gibbs function*. It is defined as

$$G = H - TS \tag{4-13}$$

where H is enthalpy, T is the absolute temperature, and S is the entropy. You can see clearly from this definition that Gibbs energy is what remains of the enthalpy after subtracting out the energy loss due to entropy.

Gibbs Energy Tells Us Something about the Spontaneity of a Process

The Gibbs energy *change* for a process is defined as

$$\Delta G = \Delta H - T\,\Delta S \tag{4-14}$$

It didn't take long for scientists to realize that if we measure the Gibbs energy change for a process that happens spontaneously, then the Gibbs energy of the final state is always less than that of the initial state. That is, for a process that happens spontaneously (i.e., on its own, with no outside push), Gibbs energy always decreases; ΔG is negative.

This was in fact what Gibbs was striving for when he proposed a way to measure the available energy. Gibbs knew well, from the branch of physics called mechanics, that a force is what moves a physical object. He also knew that there is a simple mathematical relation between force and movement ($\mathbf{F} = m\mathbf{a}$). Physical chemists at that time were looking for a property analogous to force which could be measured and which could be shown to be the force that drives chemical reactions. Gibbs found it. Just as the force of gravity causes a ball to roll down hill, so too, for any process that takes a system from state A to state B, if state B has less Gibbs energy than state A, then the process will proceed spontaneously, rolling down the Gibbs energy hill from A to B.

The Second Law of Thermodynamics

No review of thermodynamics would be complete without some discussion of the second law of thermodynamics. In all likelihood you've heard of the second law and learned something about it. And you probably know that it has something to do with entropy increasing. Perhaps you even know there are many ways to state the second law of thermodynamics, and as different as they all may sound, they are all various views of the same thing. For example,

- The universe is winding down.
- Heat spontaneously flows from a place of higher temperature to a place of lower temperature.
- When converting energy into work, some of the energy is always lost to heat.
- In any process, the entropy of a closed system will either increase or remain constant.
- In any process, the entropy of the open system and its surroundings, taken together, will either increase or remain constant.
- In any process, a system taken together with its surroundings will always go from a more ordered state to a less ordered state.

Some statements of the second law are very precise. Others, such as "The universe is winding down," can be ambiguous. Care must be taken not to misinterpret their meaning. For example, it is not scientifically correct to make statements like "His dorm room is proof of the second law of thermodynamics."

When a living thing develops from an undifferentiated mass of cells to a fully grown plant or animal, it is clear that it has gone from a less ordered state to a more ordered state. Because of this fact and the fact that living things continue to propagate their order by having offspring, scientists have speculated that living things may somehow violate the second law. Or, at least, they considered it a possibility there are physical laws we have yet to discover that are apparent only in the case of living matter.

But living things are open systems. When taken together with their surroundings, the impact of the living is to increase the disorder of the environment to a greater extent than they increase their own order. Overall the result is an increase in disorder and the second law of thermodynamics is not violated.

The Gibbs Function as the Driver of Biophysical Processes

For the most part (and this may strike you as odd) biophysics is *not* concerned with the second law of thermodynamics, at least not as stated. What concerns us more is the Gibbs function. The Gibbs function effectively has the second law of thermodynamics built into it, so by focusing on the Gibbs function alone, everything is taken care of.

In this context we can make the following statement, which is akin to other statements of the second law: All else being equal, in any process, an increase in entropy makes a favorable contribution to the Gibbs energy change, whereas a decrease in entropy makes an unfavorable contribution to the Gibbs energy change. By *favorable* and *unfavorable* we mean whether or not the process is driven forward.

So forget about the second law of thermodynamics. Well, don't actually forget about it, but focus on biophysical processes in terms of the Gibbs function being the force that drives the process.

$$\Delta G = \Delta H - T \Delta S$$

We can see from the Gibbs function that there are three things that contribute to negative ΔG and thus to rolling down the Gibbs energy hill and to driving a process forward.

1. *Releasing energy.* When a system goes from a higher-energy state to a lower-energy state, it releases energy; its enthalpy decreases. ΔH is negative, and a negative ΔH contributes toward a decrease in Gibbs energy.

2. *Increasing disorder*. When a system goes from more ordered to less ordered, its entropy increases, ΔS is positive, and so, all else being equal, its Gibbs energy decreases.

3. *Changing temperature*. Note that the effect of temperature depends on whether the entropy change is positive or negative. For a positive entropy change, increasing temperature contributes to a favorable Gibbs energy change.

Some processes are driven forward by simultaneously releasing energy and increasing disorder. For other processes it may be the case that either the enthalpy change or the entropy change are unfavorable, yet the process is still driven forward because the overall Gibbs energy change is favorable. In such a case if the process is driven forward primarily by releasing energy, then the process is said to be *enthalpy driven*. On the other hand if the process is driven forward primarily by increasing disorder, then the process is said to be *entropy driven*.

QUIZ

Refer to the text in this chapter if necessary. Answers are in the back of the book.

1. In Prob. 4-1 we made some simplifying assumptions to make it easier to calculate the change to the athlete's internal energy. Which of the following is *not* one of the simplifying assumptions that we made?

   A. We assumed we can ignore the weight of the athlete's clothing.
   B. We assumed we can ignore energy used to digest the orange juice.
   C. We assumed we can ignore energy used to light the stairway.
   D. We assumed we can ignore energy given off by the athlete's body heat.

2. The concept of *entropy* was developed

   A. to account quantitatively for the portion of energy that is not free to be converted into work.
   B. to account quantitatively for the number of different ways of doing something.
   C. by Sadi Carnot.
   D. as a measure of order in living systems.

3. What does it mean if the Gibbs energy change for a process is negative?

   A. The process is not good for living things.
   B. The process is increasing entropy.
   C. The process is not spontaneous.
   D. The process is spontaneous.

4. When a certain protein molecule binds to DNA, the entropy decreases by 2 kcal / degree. At the same time it releases 700 kcal of enthalpy. What is the Gibbs energy change for the binding reaction at a temperature of 310 K? [Hint: use Eq. (4-14)].

   A. 1320 kcal
   B. 80 kcal
   C. −80 kcal
   D. −1320 kcal

5. A mutation to the protein in question 4 results in only 600 kcal of enthalpy being released upon binding. Everything else is the same. What is the Gibbs energy change for the binding reaction with the mutated protein?

   A. 10 kcal
   B. 20 kcal
   C. 40 kcal
   D. 80 kcal

6. The concept that heat is the "lowest" form of energy means that

   A. heat will do things that no other form of energy would even think of.
   B. there is very little energy in heat.
   C. heat engines cause pollution.
   D. heat is disorganized, so only a small fraction of it can do work.

7. A "favorable" Gibbs energy change
   A. will drive a process forward.
   B. is negative.
   C. means that the Gibbs energy of the system has decreased.
   D. all of the above

8. **Which of the following statements is *not* true?**
   A. A system can do work on its surroundings.
   B. Surroundings can do work on a system.
   C. An adiabatic system is thermally isolated from its surroundings.
   D. Living organisms are adiabatic systems.

9. **The effect of changing the temperature, on the Gibbs energy change for a process, depends on**
   A. whether the process is temperature dependent.
   B. whether the entropy increases or decreases.
   C. whether the enthalpy change is favorable.
   D. none of the above

10. **A certain drug interacts by binding to neurons in the brain. The Gibbs energy change for the drug binding to a neuron is −18 kcal at body temperature (310 K). Using a calorimeter, the enthalpy change was measured to be −49 kcal. Calculate the entropy change for the drug binding to the neuron.**
    A. −0.1 kcal/deg
    B. −0.01 kcal/deg
    C. 0.15 kcal/deg
    D. −0.216 kcal/deg

chapter 5

Statistical Mechanics

In the last chapter, we learned about thermodynamics. In this chapter, we study statistical mechanics which provides a mathematical framework to explain the thermodynamics of a system at the molecular level.

CHAPTER OBJECTIVES

In this chapter, you will

- Learn the basics of statistical mechanics and how it can be used in biophysics.
- Learn the significance of the Boltzmann distribution.
- See how the partition function is used to calculate thermodynamic quantities, such as average energy.
- Use the partition function to calculate the probability of finding a molecule in a particular energy state.

What Is Statistical Mechanics?

Statistical Mechanics and Thermodynamics

Thermodynamics is the study of energy and how it operates in the physical universe. *Statistical mechanics* is the study of how probabilities and statistical averages of particles can be related to the overall thermodynamic measurements of a system. For example, the average speed (or average kinetic energy) of water molecules in a glass of water is directly related to the temperature of the water. The faster the water molecules are moving, the hotter the water.

In biophysics, the particles whose statistics we are interested in are biomolecules or, sometimes, residues and subunits within biomolecules.

Statistical mechanics provides a *molecular interpretation* to thermodynamic quantities. A molecular interpretation is an explanation of what the molecules or particles of a system are doing when the system undergoes some overall change. For example, a molecular interpretation is an answer to questions such as: What are the molecules doing when the temperature increases? What, at a molecular level, causes an enthalpy change? What happens to molecules when the entropy increases or decreases?

Statistical mechanics also places the molecular interpretation into a mathematical framework that can be used to calculate thermodynamic quantities. The calculated quantities can be used to compare with, or predict, the results of experiments.

The Process of Applying Statistical Mechanics

The first thing we do, when applying statistical mechanics, is devise a *model* that defines all of the possible microscopic energy states of the system. For example, if we are studying the protein hemoglobin which is capable of binding four oxygen molecules, we can devise a model in which the hemoglobin has five energy states. These five states might be defined by no oxygen bound, one oxygen bound, and so on, up to four oxygen molecules bound. The definition of the energy states also needs to include some information about the energy level of each state, or energy differences between each of these five energy states. All of this is fundamentally a guess or hypothesis about the system, typically based on some experimental information that we may have.

The next step is to determine the number of different ways the molecules can distribute themselves among those energy states. For example, all of the molecules could be in the highest energy state, or all of the molecules could be in in the lowest energy state, or (for our hemoglobin example) one fifth of the molecules could be in each of the five energy states. These are just three possible ways the molecules can be distributed among the energy states defined in our model. The total energy of the system is of course equal to the sum of the energy of each of the molecules. If all of the molecules are in the highest energy state then the system has the most energy that it can have. If all of the molecules are in the lowest energy state then the system has the least amount of energy that it can have. In between, depending on the total amount of energy in the system, there may be various ways to distribute the molecules among the energy levels so that the sum of their energies is still equal to the total energy of the system.

Once we have calculated all of the possible ways to distribute the molecules among the defined energy levels, we determine which of those distributions are most likely to occur. We do this using the mathematics of probability and statistics. From the distribution of energy, and the probability of each molecule being in a given energy state, we can calculate the expected value of various thermodynamic quantities. For example we could calculate the temperature of the system, or the average enthalpy change for binding an oxygen molecule. If this calculated value matches our experimental result, then we can say that our model is consistent with the data. If the calculated values do not match our experimental results, then our molecular interpretation is wrong and we have to come up with some other model or interpretation of what the molecules are doing.

Notice that when our model is *not* able to predict experimental results, this is a stronger level of proof than the case when our model *is* able to predict experimental results. When a model does *not* match experimental results, we can easily say that the model is wrong. But if our model *is* able to predict experimental results, we can't say our model is proven to be correct. It is always possible that other models, other molecular interpretations, may also be able to predict the experimental results. So at best we can only say that the model is *consistent with the data*.

Over time, if we continually design and carry out experiments striving to prove our model wrong, and those experiments fail again and again to disprove the model, then we gain confidence that our model is correct.

Sometimes a particular model can also imply results from nonthermodynamic experiments, for example, from spectroscopic or chemical studies. If such results are also consistent with the model, this further lends credence to our model being a correct interpretation of the system at a molecular level. Table 5-1 summarizes what we can conclude from experiments depending on whether or not the model's predictions are consistent with the experimental results.

TABLE 5-1 Summary of relationship between model and experiment, and what we can conclude on the basis of that relationship.

Relationship between Model and Experiment	What We Can Conclude
Model is NOT able to correctly predict experimental results.	The model is definitely wrong in some way. Back to the drawing board. Hopefully only slight changes are needed.
Model accurately predicts the results of experiments.	The model is consistent with the data. The model may be correct. More experimentation is needed. What else can the model predict? How can we try to prove the model wrong?

Statistical Mechanics: A Simple Example

Let's take a look at an example. We start with a very simple system, only four molecules. Suppose that each of these four molecules can only have an amount of energy equal to some multiple of 10^{-20} J, for example, 1×10^{-20} J, 2×10^{-20} J, 3×10^{-20} J, and so on. We will call 10^{-20} J our unit of energy for purposes of this example.

Our intention is only to illustrate how statistical mechanics works. So it's not important *why* the molecules are limited to these specific energy values. In a real situation, however, the model would typically define why there are limits on the possible energy states of molecule. Such specifics can add to the model's ability to predict experimental results. For now, just take it as a given that our molecules can only be in certain, defined energy states, and later we will give some biophysical examples of how this can happen.

Let's say that our system of four molecules has a total energy of 6×10^{-20} J. We assume each molecule must have at least 1×10^{-20} J of energy and, as stated, each molecule can only have values of energy that are integer multiples of 10^{-20} J. With these assumptions there are only *two* ways to distribute 6×10^{-20} J of energy among the four molecules. These are shown in Fig. 5-1.

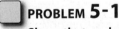 PROBLEM 5-1

Show that each distribution in Fig. 5-1 has a total energy of 6×10^{-20} J.

 SOLUTION

In Fig. 5-1a, three molecules have 1×10^{-20} J per molecule, and one molecule has 3×10^{-20} J, so the total energy is (1 molecule $\times$ 3×10^{-20} J/molecule) + (3 molecules $\times$ 1×10^{-20} J/molecule) = 6×10^{-20} J.

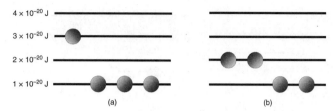

FIGURE 5-1 • There are only two ways to distribute 6×10^{-20} J of energy among four molecules, given the constraints that (1) each molecule must have at least 1×10^{-20} J of energy and (2) each molecule can only have an amount of energy that is an integer multiples of 10^{-20} J. (a) Three molecules have 1×10^{-20} J each, and one molecule has 3×10^{-20} J. (b) Two molecules have 2×10^{-20} J and two have 1×10^{-20} J.

In Fig. 5-1b, two molecules have 2×10^{-20} J per molecule and two have 1×10^{-20} J per molecule, so the total energy is (2 molecules $\times$ 2 $\times$ 10^{-20} J/molecule) + (2 molecules $\times$ 1 $\times$ 10^{-20} J/molecule) = 6 $\times$ 10^{-20} J.

Permutations

Each of the distributions shown in Fig. 5-1 has several ways to arrange the four molecules to still achieve the same distribution. For example, in the case of distribution (a), the distribution is defined by having three molecules with 1 unit (10^{-20} J) of energy and one molecule with 3 units of energy. Any one of the four molecules could be the one with 3 units of energy, so there are four different ways to arrange the molecules to achieve the same distribution. These arrangements or *permutations* are shown in Fig. 5-2.

In the case of the distribution shown in Fig. 5-1b, any two of the four molecules could be the two molecules with 2×10^{-20} J. There are six different ways to choose two items out of the four. So there are six permutations or arrangements in which two of the four molecules will have 2 units of energy and two will have 1 unit of energy. These six permutations are shown in Fig. 5-3.

Altogether in Figs. 5-2 and 5-3 there are ten different ways to partition the 6 units of energy among the four molecules, assuming each molecule must have

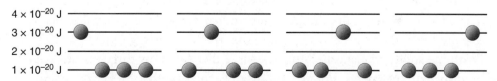

FIGURE 5-2 • Four different ways to achieve the energy distribution shown in Fig. 5-1a.

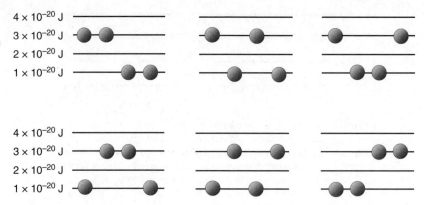

FIGURE 5-3 • Six different ways to achieve the energy distribution shown in Fig. 5-1b.

at least 1 unit of energy. Four of the partitions correspond to the energy distribution shown in Fig. 5-1a, and six of them correspond to the distribution shown in Fig. 5-1b. If we assume that each of these ten partitions is equally likely to occur (and there is no reason not to), then at any given moment in time there is a 40% probability that the energy will be distributed according to Fig. 5-1a and a 60% probability the energy will be distributed according to Fig. 5-1b.

In this example, distribution b is called the *most probable distribution* because it has the largest number of ways to arrange the molecules, that is, the largest number of permutations. In general, we identify the most probable distribution as the distribution with the largest number of permutations. We can do this because we have assumed that all arrangements (permutations) are equally likely, so the distribution with the most permutations will be most likely to occur.

In this specific example it turns out that the other distribution, the one that is *not* the most probable, also has a significant probability of occurring (40% in our example). This is not always the case. We will soon see that as we increase the number of molecules, the most probable distribution rapidly begins to overshadow all other distributions, to the point where the probability of finding the system in any other distribution is very small compared to the most probable distribution. So small in fact that it becomes *negligible;* this means we can safely ignore all but the most probable distribution and the results of any calculations will be essentially the same as if we had included those distributions.

Before we add molecules to our system, let's see what happens if we add energy to our system. Suppose we heat up our system of four molecules, increasing the temperature ever so slightly, so that the total energy is now 7×10^{-20} J. With the addition of another unit of energy there are now three possible ways to distribute the energy among the four molecules (see Fig. 5-4).

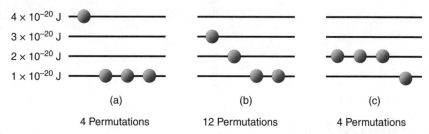

4×10^{-20} J

3×10^{-20} J

2×10^{-20} J

1×10^{-20} J

(a)   (b)   (c)

4 Permutations   12 Permutations   4 Permutations

FIGURE 5-4 • There are three ways to distribute 7×10^{-20} J of energy among four molecules (given the same two constraints as in Fig. 5-1). The three distributions, as shown in the figure, are: (a) Three molecules have 1×10^{-20} J each, and one molecule has 4×10^{-20} J. (b) Two molecules have 1×10^{-20} J each, one molecule has 2×10^{-20} J, and one molecule has 3×10^{-20} J. (c) One molecule has 1×10^{-20} J, and three molecules have 2×10^{-20} J. The figure also shows, underneath each distribution, the number of permutations for that distribution.

Increasing the total energy of the system has increased the number of energy levels *available* to the molecules. Previously, because each molecule must have at least 1 unit of energy, the highest energy level any one molecule could have was 3×10^{-20} J (since the other 3×10^{-20} J had to be distributed among the other three molecules to ensure that each one has at least 1×10^{-20} J). Therefore there were a total of three energy levels available to any of the molecules: 1×10^{-20} J, 2×10^{-20} J, and 3×10^{-20} J.

Now with a total energy of 7×10^{-20} J, there are four energy levels available. The additional energy level (4×10^{-20} J) has increased the total number of ways to arrange the molecules from 10 ways to 20 ways for the same four molecules. Notice that one particular distribution, Fig. 5-4b, still stands out as having the greatest number of permutations and is therefore the most probable distribution.

There is a simple formula for calculating the number of different ways to arrange the molecules in any given distribution. Given a total of L_{MAX} energy levels, a distribution is defined by specifying how many molecules are at each energy level. That is, we define a distribution by saying there are n_1 molecules at energy level L_1, n_2 at level L_2, and so on, up to $n_{L_{MAX}}$. If the total number of molecules is N, the number of different ways of arranging N molecules into L_{MAX} groups, with n_1 in group 1, n_2 in group 2, . . . , is

$$W = \frac{N!}{n_1! \cdot n_2! \cdot n_3! \dots n_{L_{MAX}}!}$$ (5-1a)

Just as a reminder, we note that the symbol ! stands for factorial. For example, the term $n_1!$ (n_1 factorial) means $n_1 \cdot (n_1 - 1) \cdot \cdots \cdot 2 \cdot 1$. Of course, if $n_1 = 2$, then $n_1! = 2 \cdot 1 = 2$, and if $n_1 = 1$, then $n_1! = 1$. Zero factorial is defined as equal to 1, so if any $n_i = 0$, then we use 1 for $n_i!$ in Eq. (5-1a).

Note that the number of molecules N is equal to the sum of all the n_i from $i = 1$ to $i = L_{MAX}$.

$$N = \sum_{i=1}^{L_{MAX}} n_i \qquad (5\text{-}2)$$

 PROBLEM 5-2

Use Eq. (5-1a) to verify that the number of permutations listed for each of the distributions in Fig. 5-4 is correct.

 SOLUTION

For distribution a we have a total of four energy levels ($L_{MAX} = 4$). There are three molecules on level 1 and one molecule on level 4. There are zero molecules on levels 2 and 3, so

$$n_1 = 3, \quad n_2 = 0, \quad n_3 = 0, \quad \text{and} \quad n_4 = 1$$

Therefore, per Eq. (5-1a),

$$W = \frac{4!}{3! \cdot 0! \cdot 0! \cdot 1!} = \frac{4 \cdot 3 \cdot 2 \cdot 1}{(3 \cdot 2 \cdot 1) \cdot (1) \cdot (1) \cdot (1)} = 4$$

For distribution b,

$$n_1 = 2, \quad n_2 = 1, \quad n_3 = 1, \quad \text{and} \quad n_4 = 0$$

so

$$W = \frac{4!}{2! \cdot 1! \cdot 1! \cdot 0!} = \frac{4 \cdot 3 \cdot 2 \cdot 1}{(2 \cdot 1) \cdot (1) \cdot (1) \cdot (1)} = 4 \cdot 3 = 12$$

For distribution c,

$$n_1 = 1, \quad n_2 = 3, \quad n_3 = n_4 = 0$$

so

$$W = \frac{4!}{1! \cdot 3! \cdot 0! \cdot 0!} = \frac{4 \cdot 3 \cdot 2 \cdot 1}{(1) \cdot (3 \cdot 2 \cdot 1) \cdot (1) \cdot (1)} = 4$$

Increasing the Number of Molecules

Next let's see what happens if we increase both the total energy and the total number of molecules in the system. We'll start out keeping things relatively simple, but as we will see adding just a few molecules very quickly increases the number of ways, we can partition energy in the system. Consider the case of only ten molecules with a total of 18 units of energy.

PROBLEM 5-3

Determine how many energy levels are available to ten molecules with a total of 18 units of energy among them. Assume that each molecule must have at least 1 unit of energy and that the spacing between successive energy levels is uniform with 1 unit of energy between each level.

SOLUTION

We find the number of energy levels by finding the maximum energy that any one molecule can have. To do this, first give the minimum amount of energy to each molecule, for all but one molecule. This effectively sets aside the most amount of energy that can be left over for one molecule.

The constraints of the problem state that each molecule must have at least 1 unit of energy. There are ten molecules, so first give nine of the ten molecules 1 unit of energy each. From the 18 original units of energy, we now have 9 units of energy remaining to give to the last molecule. Therefore the most energy any one molecule can have is 9 units. This means there are a total of nine energy levels available to ten molecules with 18 units of energy total.

Finding Distributions

The way we distributed the energy in Sol. 5-3, by giving the minimum amount of energy to all but one molecule and then giving all of the remaining energy to the last molecule, is a good starting point for finding all possible distributions. It is the simplest distribution possible. It is the only distribution in which a single molecule has more than the minimum amount of energy. And that one molecule will always occupy the highest energy level available to any of the molecules.

It is also the distribution with the smallest number of permutations (ways to arrange the molecules within the same distribution). The number of permutations for this distribution is simple to determine. Any one of the molecules could be the one molecule at this highest available energy level. (The remaining molecules are all on the lowest energy level.) Therefore the number of permutations for this simplest distribution is exactly equal to the number of molecules, as each molecule takes its turn being the one molecule at the highest energy level.

Once we have defined this simplest distribution, every other distribution can be found by taking one or more units of energy from the molecule with the most energy and redistributing that energy to one or more of the remaining molecules.

In our present example then, we take 1 or more units of energy from the molecule with 9 units of energy and redistribute that energy to one or more of the remaining nine molecules. For ten molecules with 18 units of energy there are 22 possible distributions. Table 5-2 lists all 22 possible distributions of 18 units of energy among ten molecules, given the constraint that each molecule

TABLE 5-2 Distributions for ten molecules with 18 units of energy and number of permutations for each.

Distribution	Permutations
9 1 1 1 1 1 1 1 1 1	10
8 2 1 1 1 1 1 1 1 1	90
7 3 1 1 1 1 1 1 1 1	90
7 2 2 1 1 1 1 1 1 1	360
6 4 1 1 1 1 1 1 1 1	90
6 3 2 1 1 1 1 1 1 1	720
6 2 2 2 1 1 1 1 1 1	840
5 5 1 1 1 1 1 1 1 1	45
5 4 2 1 1 1 1 1 1 1	720
5 3 3 1 1 1 1 1 1 1	360
5 3 2 2 1 1 1 1 1 1	2520
5 2 2 2 2 1 1 1 1 1	1260
4 4 3 1 1 1 1 1 1 1	360
4 4 2 2 1 1 1 1 1 1	1260
4 3 3 2 1 1 1 1 1 1	2520
4 3 2 2 2 1 1 1 1 1	5040
4 2 2 2 2 2 1 1 1 1	1260
3 3 3 3 1 1 1 1 1 1	210
3 3 3 2 2 1 1 1 1 1	2520
3 3 2 2 2 2 1 1 1 1	3150
3 2 2 2 2 2 2 1 1 1	840
2 2 2 2 2 2 2 2 1 1	45
Total Permutations for All Distributions Combined	**24310**

must have at least 1 unit of energy. In the table, the first ten columns represent each of the ten molecules, and the number in the column indicates how many units of energy that molecule has. The last column shows the number of permutations for each distribution.

Each row shows one of 22 possible distributions. Each distribution is represented by a list of 10 numbers, one number for each molecule. The number value is how much energy that molecule has (or, in other words, which energy level that molecule is on).

Still Struggling

There are two equivalent ways to speak about statistical mechanical distributions. We can speak about distributing energy among molecules, or we can speak about distributing molecules among energy levels. These are the same; they are just two ways of looking at the same situation.

In our present example, when distributing *energy among molecules,* we say there are 22 possible distributions of 18 units of energy among ten molecules, given the constraint that each molecule must have at least 1 unit of energy. In this way of speaking, then, we say the number in each molecule column of Table 5-2 indicates how many units of energy that molecule has.

When speaking about distributing *molecules among energy levels*, we say there are nine energy levels available to the molecules, and there are 22 possible distributions of ten molecules among nine energy levels, given the constraint that the total energy must always add up to 18 units. In Table 5-2 then, we would say the number in each molecule column indicates which energy level that molecule occupies.

Ignoring Distributions

We should emphasize again that all *permutations,* or all ways of partitioning the energy, are equally likely regardless of which *distribution* they belong to. This means that if we have a distribution A with 100 permutations, and a distribution B with only 5 permutations, then at any given point in time we are 20 times more likely to find our system in distribution A than in distribution B.

Let's take a closer look at the case of ten molecules with 18 units of energy among them. Seven of the 22 possible distributions are illustrated in Fig. 5-5.

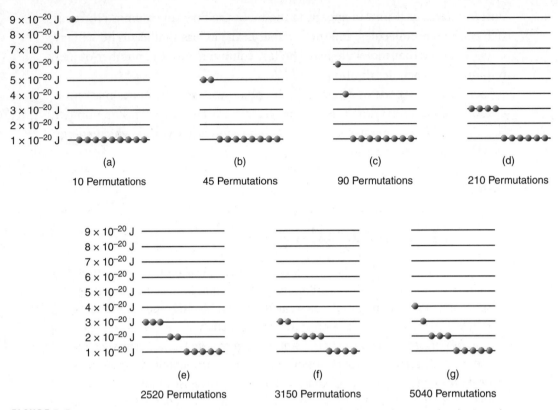

FIGURE 5-5 • Diagram of seven of the 22 possible distributions listed in Table 5-2. The seven distributions shown include the three distributions with the largest number of permutations and four distributions from among those with the fewest number of permutations.

For comparison purposes, we have chosen the three most probable distributions and four of the least likely distributions.

Looking at the four least likely distributions in Fig. 5-5, the total number of ways to arrive at any of these four distributions is the sum of their permutations, or 355. If we compare this to the total number of permutations of all 22 distributions (i.e., 355 compared to 24,310), we see that these four distributions combined amount to less than 1.5% of the total number of ways to partition the energy. That is, there is less than a 1.5% chance of finding our system in any of these four distributions combined. This suggests that we could simply ignore these four distributions, and we would introduce an error of no more than 1.5% to our calculations. Such a small amount of error is insignificant for many situations. We can therefore say that in such situations these distributions are *negligible;* that is, they can safely be ignored.

Looking at all of the distributions in Table 5-2, if we sort them by the number of permutations and start with those distributions that are least likely to occur (those with the fewest number of permutations), we can actually ignore 9 of the 22 distributions and introduce only about a 5% error. We could even ignore half of the distributions and introduce less than a 10% error.

We can visualize which distributions can be safely ignored in the following way: We sort the distributions from those with the least number of permutations to those with the most. We assign each distribution an identification number from 1 to N_D, where N_D is the total number of distributions. For any given distribution number x, we calculate the cumulative probability for all of the distributions up to and including that distribution. This is the same as the combined number of permutations divided by the total number of permutations for all distributions. We then plot this cumulative probability on the y-axis, against the distribution number x on the x-axis (where x runs from 1 to N_D). The result is shown in Fig. 5-6a, where you can clearly see that nearly half (10 of 22) of the distributions contribute a *cumulative* (combined) probability of no more than .14 and so half the distributions can be ignored if we are willing to tolerate a 14% error in our calculations.

Once we begin to deal with large numbers of molecules and even larger numbers of distributions, it is convenient to plot the percentage of distributions on the x-axis, instead of the distribution number. This is shown in Fig. 5-6b where it

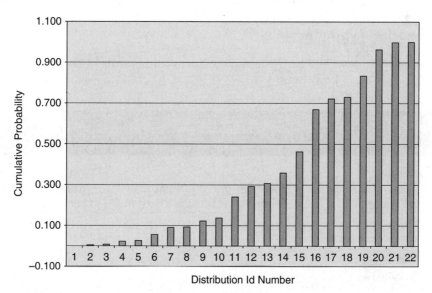

FIGURE 5-6A • Cumulative probability for the 22 possible distributions for a system of ten molecules with 18 units of energy. The distributions are plotted in order of increasing number of permutations, so that it is easy to see how many distributions can be ignored while introducing only minimal error.

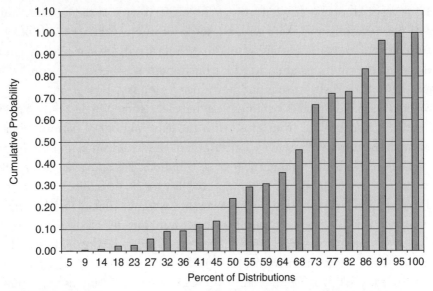

FIGURE 5-6B · Cumulative probability as a function of the percentage of distributions that contribute to that probability. It is easy to see that we can ignore 45% of the distributions and have no more than a 15% probability that the system would have been in one of those distributions.

is simple to see that if we want to ignore distributions that contribute a combined probability of no more than .15, then we can ignore 45% of the distributions.

Comparing Examples

Table 5-3 compares the case of ten molecules and 18 units of energy with our previous two examples. Notice that with only a modest increase from our first

TABLE 5-3 Comparison of number of distributions and permutations for three different statistical mechanical models.

Example: Number of Molecules Total Energy	Number of Energy Levels	Number of Distributions	Total Number of Arrangements (Permutations)
4 Molecules 6×10^{-20} J (6 units)	3	2	10
4 Molecules 7×10^{-20} J (7 units)	4	3	20
10 Molecules 18×10^{-20} J (18 units)	9	22	24,310

case, six additional molecules and six additional available energy levels, the result is more than 10 times as many distributions and more than 2000 times as many total ways to arrange the molecules! We see from Table 5-3 that things can become very complex very quickly by the addition of only a few molecules.

Table 5-3 compares three simple statistical mechanical examples: Four molecules with 6 units of energy to distribute among them, four molecules with 7 units of energy, and ten molecules with 18 units of energy. Notice that increasing the amount of energy or the number of molecules, even by a small amount, results in a significant increase in the number of possible distributions as well as the total number of permutations among all distributions.

Now let's see what happens if we increase the number of molecules further. Until now, we have kept the number of molecules very small in order to better illustrate the principles of statistical mechanics. In a real-life biophysical situation, when applying statistical mechanics, we would typically have at least a thousand residues, if not billions of molecules, even up to the order of 10^{20} molecules or more. Keep these numbers in mind as we discuss the next several examples, which will still be relatively simple (less than 100 molecules) but which will help illustrate the principles that come into play when we deal with large numbers of molecules.

In the examples that follow, as we increase the number of molecules, we keep the average energy per molecule constant at 1.8×10^{-20} J. This places the examples on equal footing for comparing just the effect of increasing the number of molecules.

Table 5-4 shows our previous example, 10 molecules with 18 units of energy, along with additional examples of 5 molecules with 9 units of energy, 20 molecules with 36 units of energy, 30 molecules with 54 units, and so on. (In all cases the average energy is 1.8 units per molecule.) In going from 10 molecules to 50 molecules we have increased the number of molecules by 5 times, yet the total number of permutations increases by 10^{19} times! You can see that increasing the number of molecules just a little (50 is still a very small number of molecules) already results in an unwieldy number of molecular arrangements to consider.

There is in Table 5-4, however, another trend that can help us here. The trend is that, as we increase the number of molecules, the vast majority of the molecular arrangements are found in only a very small percentage of the distributions.

TABLE 5-4   Effect of number of molecules on the number of distributions, on the total number of permutations, and on the number of distributions that can be ignored.

Number of Molecules	Number of Distributions	Total Number of Permutations	Percent of Distributions That Can Be Ignored with Less than 5% Error
5	5	9	14.0%
10	22	24,310	37.0%
20	231	4.1×10^9	71.0%
30	1,575	7.8×10^{14}	87.0%
40	8,349	1.6×10^{20}	94.0%
50	37,338	3.3×10^{25}	97.2%
60	147,273	7.1×10^{30}	98.7%
70	526,823	1.5×10^{36}	99.4%
80	1,741,630	3.4×10^{41}	99.7%
90	5,392,783	7.5×10^{46}	99.9%
100	15,796,476	1.7×10^{52}	99.9%

Take a look at the last column in Table 5-4. This column shows the percentage of distributions that can be safely ignored because their cumulative (combined) probability is less than 5% of the total. By the time we reach 60 molecules, we can ignore 99% of the distributions. At 90 molecules we can ignore 99.9% of the distributions. You can see that when dealing with a more realistic number of molecules (thousands, millions, billions, etc.), *we can easily ignore all but an extremely small percentage of distributions* (in fact, as we discuss below, we can ignore all but the one most probable distribution).

This trend is depicted rather vividly in Figs. 5-7 through 5-9. You should compare these figures with Fig. 5-6b in which we plotted the cumulative probability as a function of the percentage of distributions that contribute to that cumulative probability for a system of ten molecules. Figs. 5-7 through 5-9 plot the same for systems of 20, 30, and 40 molecules, respectively. The figures show very clearly that as we increase the number of molecules, the vast majority of distributions have a very low probability of occurrence.

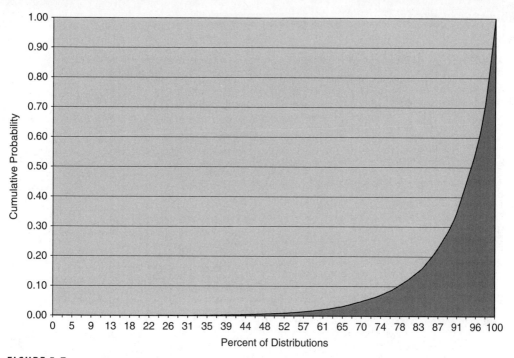

FIGURE 5-7 · Cumulative probability versus percent of distributions contributing to that probability for a system of 20 molecules.

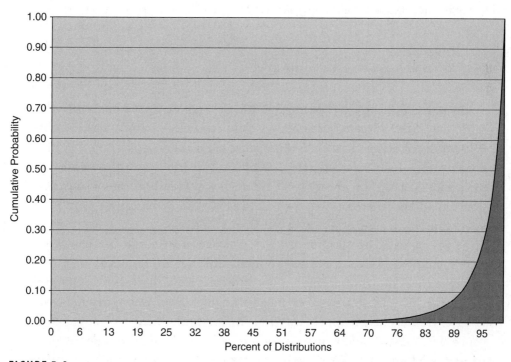

FIGURE 5-8 · Cumulative probability versus percent of distributions contributing to that probability for a system of 30 molecules.

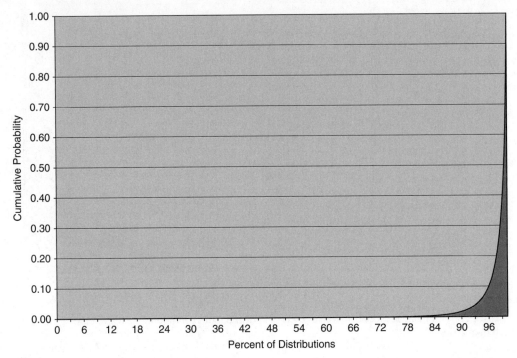

FIGURE 5-9 • Cumulative probability versus percent of distributions contributing to that probability for a system of 40 molecules.

Once we have enough molecules so to safely ignore 99.9% of the distributions, there is another fact that allows us to ignore even more. The less than 0.1% of distributions that we do not ignore are all very similar to each other. For example, for this small set of significant distributions, the number of molecules on any given energy level differs by less than 1% from one distribution to another. Therefore any one of these distributions can be used as an approximation for the others with less than 1% error.

As we increase the number of molecules even further, the percentage of significant distributions becomes so small and the differences among them so slight that the whole problem becomes simply a matter of finding the most probable distribution. The most probable distribution, in such cases, is an extremely close approximation for any distribution of any significance. The most probable distribution is called the *Boltzmann distribution*, named for Ludwig Boltzmann who first derived a formula for it in the 1860s.

Still Struggling

Many texts, when discussing statistical mechanics, will speak of the Boltz-mann distribution as if it is the only distribution found in nature, that is, the only way that nature distributes energy within a system of particles. You should keep in mind that other distributions can and do occur. However, their probability is typically very low and their existence so transitory that for any realistic and significant number of molecules, the Boltzmann distri-bution is all that matters. Only in the extreme case of small numbers of mol-ecules and/or relatively little energy, do we need to consider other distributions in our calculations.

Boltzmann Distribution

Finding the most probable distribution is just a matter of finding the distribution with the largest number of ways of arranging the molecules.

As before, we define a distribution (of molecules among energy levels) by specifying the number of molecules at each available energy level. If we write the number of molecules on energy level i as n_i, then a distribution is defined by specifying all n_i from n_1 through $n_{L_{MAX}}$, where L_{MAX} is the highest energy level available to the system.

If we can come up with a formula for n_i for the most probable distribution, we can use that formula in all of our statistical mechanical calculations, and this is exactly what Boltzmann did.

One more thing to note about L_{MAX}: Unless otherwise stated, we assume that all energy levels between 1 and L_{MAX} are available to the molecules and L_{MAX} is not just the highest available energy level but is also equal to the total number of available energy levels.

Once we have any distribution, that is, once we know all of the n_i, then regardless of what that distribution is, the following two relationships apply:

The total number of molecules is equal to the sum, over all of the energy levels, of the number of molecules on each energy level, as we saw in Eq. (5-2).

$$N = \sum_{i=1}^{L_{MAX}} n_i$$

The number of different ways to arrange the molecules and still have the same distribution is given by Eq. (5-1a). We can rewrite Eq. (5-1a) using the product symbol (π) meaning take the product of all $n_i!$ from $i = 1$ to L_{MAX}.

$$W = \frac{N!}{\prod\limits_{i=1}^{L_{MAX}}(n_i!)} \qquad (5\text{-}1b)$$

Mathematically, finding the most probable distribution amounts to finding the set of n_i that maximizes Eq. (5-1b) for a given value of N and of L_{MAX}. We will skip the somewhat involved mathematics for maximizing Eq. (5-1b). The interested reader can look up the derivation in most libraries or on the Web. The end result is that the set of n_i that maximizes W is given by

$$n_i = \frac{N}{Z}e^{-\beta E_i} \qquad (5\text{-}3)$$

where n_i is the number of molecules with energy E_i (i.e., on energy level i), N is the total number of molecules, β is a constant that can be determined by comparison with experimental data, and Z is the sum of all $e^{-\beta E_i}$ for all energy levels; that is,

$$Z = \sum\limits_{i=1}^{L_{MAX}} e^{-\beta E_i} \qquad (5\text{-}4)$$

Z is called the *partition function*. It contains all of the information as to how the total energy is partitioned among the molecules. Comparison with experimental thermodynamic data shows the $\beta = 1/kT$ where T is the absolute temperature and k is *Boltzmann's constant*, equal to 1.3806×10^{-23} J/K. Thus the classic formula for the Boltzmann distribution of particles among energy levels is

$$n_i = \frac{N}{Z}e^{-\frac{E_i}{kT}} \qquad (5\text{-}5)$$

Things to remember about the Boltzmann distribution are

1. It is simply the most probable distribution of N molecules across L_{MAX} energy levels. We call it the *Boltzmann distribution* (instead of the most probable distribution) to honor Ludwig Boltzmann who first derived a formula for it in the 1860s.

2. For any reasonable number of molecules and amount of energy, the probability of the Boltzmann distribution overshadows all other distributions.

This does not mean that the other distributions do not occur, only that they occur with such low probability and typically so transiently that for nearly all intents and purposes calculations based on the Boltzmann distribution are accurate and precise. We need not worry about other distributions except in the extreme case of a small number of molecules or a very small amount of energy to distribute.

3. Equation (5-5) assumes that N is at least 60 or so. This is because it uses *Stirling's approximation* for n factorial. However, there are times when even for $N < 60$, Boltzmann's distribution gives a close enough approximation.

Statistical Mechanical Calculations

Now that we have only a single distribution to worry about, we can use that distribution to make statistical mechanical calculations for

- The fraction of molecules in a particular energy state
- The average and total energy of the system

The fraction, or proportion, of molecules on a particular energy level i is the number of molecules on that energy level divided by the total number of molecules.

$$F_i = \frac{n_i}{N} = \frac{e^{-\frac{E_i}{kT}}}{Z} \tag{5-6}$$

This is also the probability of finding that any particular molecule has energy E_i. For example, if 1/5 of the molecules have energy E_i and if we could somehow sample and measure the energy of a single molecule, there would be a probability of 1/5, or .2 of finding that the molecule we sampled had an energy of E_i. Therefore we sometimes write P_i instead of F_i.

$$P_i = \frac{n_i}{N} = \frac{e^{-\frac{E_i}{kT}}}{Z} \tag{5-7}$$

The point is that the fraction of molecules in a particular state is equivalent to the probability of finding a molecule in that particular state. This is an important point because sometimes we want to know the probability of a particular biophysical process occurring. The probability of a process occurring is often related to the probability of a particular energy state occurring.

Still Struggling?

How can the probability of a biophysical process be related to the probability of a particular energy state? Let's take an example. Suppose the body needs more insulin. In order for our cells to make more insulin, the gene that contains instructions for making insulin needs to be expressed. This means that a certain group of proteins needs to bind to the DNA at a point on the DNA very near to where the beginning of that gene is. This in turn may require the DNA to be in a particular conformation, or to be supercoiled to a certain extent, thus defining a specific energy level that the DNA must be in for insulin to be made. The probability of finding the DNA in that state affects the probability of insulin being produced. Similarly the proteins involved may also each need to be in a particular conformation in order to bind to the DNA, thus defining a particular energy state for each protein. The probabilities of each of these states, for the DNA and for the proteins, affect the probability of insulin being produced. The body adjusts conditions as needed in order to alter these probabilities: increasing the probability when insulin is needed, and decreasing the probability when not. Statistical mechanics can be used to calculate these probabilities under various conditions and thus help us to understand the mechanisms involved at a molecular level.

The average energy (per molecule) in the system is the amount of energy at each energy level times the fraction of molecules at that particular energy level, summed over all possible energy levels.

$$\langle E \rangle = \sum_{i=1}^{L_{MAX}} E_i F_i = \sum E_i \frac{n_i}{N} = \sum E_i \frac{e^{-\frac{E_i}{kT}}}{Z} \tag{5-8}$$

PROBLEM 5-4

Let's say we have a sample containing a very large number of a particular biological molecule and we would like to model the various conformational states of this molecule as having three possible energy levels. In our model, the three energy levels available to our molecule are 4.0×10^{-20} J, 4.4×10^{-20} J, and 4.8×10^{-20} J. Assuming the Boltzmann distribution, calculate the fraction of molecules expected to be at each energy level at room temperature (295 K).

SOLUTION

You'll probably want your calculator for this one. Typically the first thing we do in statistical mechanics is calculate the partition function. Everything else we might want to calculate depends on the partition function. The partition function in our case is given by

$$Z = \sum_{i=1}^{L_{MAX}} e^{-\frac{E_i}{kT}} = e^{-\frac{4.0 \times 10^{-20}\,J}{kT}} + e^{-\frac{4.4 \times 10^{-20}\,J}{kT}} + e^{-\frac{4.8 \times 10^{-20}\,J}{kT}}$$

Since the energy levels are expressed in joules, we express Boltzmann's constant in joules: $k = 1.3806 \times 10^{-23}$ J/K. Then the first term of the partition function is

$$e^{-\frac{4.0 \times 10^{-20}\,J}{kT}} = e^{-\frac{4.0 \times 10^{-20}\,J}{(1.3806 \times 10^{-23}\,J/K)(295\,K)}} = e^{-9.8213} = 5.4282 \times 10^{-5}$$

Similarly the second term is

$$e^{-\frac{4.4 \times 10^{-20}\,J}{(1.3806 \times 10^{-23}\,J/K)(295\,K)}} = e^{-10.803} = 2.0329 \times 10^{-5}$$

And the third term is

$$e^{-\frac{4.8 \times 10^{-20}\,J}{(1.3806 \times 10^{-23}\,J/K)(295\,K)}} = e^{-11.786} = 0.76135 \times 10^{-5}$$

So the total value of the partition function at 295 K is

$$5.4242 \times 10^{-5} + 2.0329 \times 10^{-5} + 0.76135 \times 10^{-5} = 8.2224 \times 10^{-5}$$

Notice that all of the units cancel out. The partition function is a dimensionless quantity.

Now that we have calculated the partition function, the fraction of molecules at each energy level is given by Eq. (5-6).

$$F_1 = \frac{e^{-\frac{E_i}{kT}}}{Z} = \frac{e^{-\frac{4.0 \times 10^{-20}\,J}{(1.3806 \times 10^{-23}\,J/K)(295\,K)}}}{Z} = \frac{5.4282 \times 10^{-5}}{8.2224 \times 10^{-5}} = 66.0\%$$

Similarly,

$$F_2 = \frac{2.0329 \times 10^{-5}}{8.2224 \times 10^{-5}} = 24.7\%$$

and

$$F_3 = \frac{0.76135 \times 10^{-5}}{8.2224 \times 10^{-5}} = 9.3\%$$

PROBLEM 5-5

Given the same model as in Prob. 5-4, (three energy levels available: 4.0×10^{-20} J, 4.4×10^{-20} J, and 4.8×10^{-20} J, and the Boltzmann distribution) calculate the fraction of molecules expected to be at each energy level at body temperature (310 K).

SOLUTION

Everything is the same as in Prob. 5-4, except now the temperature is 310 K instead of 295 K. The three terms of the partition function are

$$e^{-\frac{4.0 \times 10^{-20} \text{ J}}{(1.3806 \times 10^{-23} \text{ J/K})(310\text{K})}} = e^{-9.3461} = 8.7305 \times 10^{-5}$$

$$e^{-\frac{4.4 \times 10^{-20} \text{ J}}{(1.3806 \times 10^{-23} \text{ J/K})(310\text{K})}} = e^{-10.281} = 3.4288 \times 10^{-5}$$

$$e^{-\frac{4.8 \times 10^{-20} \text{ J}}{(1.3806 \times 10^{-23} \text{ J/K})(310\text{K})}} = e^{-11.215} = 1.3466 \times 10^{-5}$$

The fraction of molecules at each energy level is

$$F_1 = \frac{e^{-\frac{4.0 \times 10^{-20} \text{ J}}{(1.3806 \times 10^{-23} \text{ J/K})(310\text{K})}}}{Z} = \frac{8.7305 \times 10^{-5}}{13.506 \times 10^{-5}} = 64.6\%$$

$$F_2 = \frac{3.4288 \times 10^{-5}}{13.506 \times 10^{-5}} = 25.4\%$$

$$F_3 = \frac{1.3466 \times 10^{-5}}{13.506 \times 10^{-5}} = 10.0\%$$

Compare this with the results from Prob. 5-4. The increase in temperature from 295 K to 310 K causes a *decrease* in the fraction of molecules at the lowest energy level, from 66% to 64.6%. At the same time it causes an *increase* in the fraction of molecules in the higher energy levels. F_2 went from 24.7% to 25.4%, and F_3 went from 9.3% to 10.0%. These results are summarized in Table 5-5. Increasing the temperature increases the internal energy of the system which, according to statistical mechanics, is equivalent to increasing the average energy of the molecules in the system. Statistical mechanics also tells us that this increase in average energy comes about by moving some of the molecules in the lowest energy level to higher energy levels.

TABLE 5-5 Fraction of molecules at each energy level for 295 K and 310 K.		
	T = 295 K	*T* = 310 K
F_1	66.0%	64.6%
F_2	24.7%	25.4%
F_3	9.3%	10.0%

Degeneracy

In statistical mechanics we speak of atoms or molecules existing in different states, where each state corresponds to some change in the atom or molecule. A state could be a particular confirmation, a particular mode of vibration, or a particular configuration of the electrons. In many cases, each state is on its own a unique energy level.

Sometimes, however, two or more states have the same energy. The molecule can be in either of these two states and still be on the same energy level. When this happens, we say that the energy level is *degenerate*, meaning that there is more than one state with that amount of energy. The *degeneracy* of the energy level is the *number* of different states with that energy.

For example, suppose a molecule has seven different conformations (shapes) that it can be in. Now suppose of the seven different conformational states, one shape has energy E_1, one has energy E_2, *four* have energy E_3, and one has energy E_4. We say that energy level 3 is *degenerate* because there is more than one confirmation with energy E_3. The *degeneracy* of energy level E_3 is 4. There are four states all with energy E_3.

There are two ways of dealing with degeneracy in statistical mechanics. One way is to sum over all possible states of the system. So, for example, we write the partition function

$$Z = \sum_{j=1}^{N_{STATES}} e^{-\frac{E_j}{kT}}$$

(5-9)

N_{STATES} is the number of states available to each molecule in the system. Each j corresponds to a different state of the system, but the E_j are not necessarily all different; two or more of the E_j can be the same. If there is no degeneracy, then all of the E_j are different, and N_{STATES} is equal to L_{MAX}. But if there is degeneracy, then at least some of the E_j will be the same and N_{STATES} will be larger than L_{MAX}.

The second way to deal with degeneracy is to sum over energy levels (as we did previously) but to multiply each term of the partition function by the degeneracy of the energy level corresponding to that term. Thus the partition function would be written

$$Z = \sum_{i=1}^{L_{MAX}} \omega_i e^{-\frac{E_i}{kT}} \qquad (5\text{-}10)$$

where ω_i is the degeneracy of the ith energy level.

The two ways of dealing with degeneracy are equivalent. The method we choose will depend on which is easier to use or more beneficial for the particular situation. Summing over energy levels is mathematically simpler, so we will use that method most commonly unless there is a specific benefit to summing over states. In the case of the example of a molecule with seven conformational states, the difference between the two methods is the difference between the following two equations:

Summing over states,

$$Z = e^{\frac{-E_1}{kT}} + e^{\frac{-E_2}{kT}} + e^{\frac{-E_3}{kT}} + e^{\frac{-E_3}{kT}} + e^{\frac{-E_3}{kT}} + e^{\frac{-E_3}{kT}} + e^{\frac{-E_4}{kT}} \qquad (5\text{-}11)$$

Summing over energy levels,

$$Z = e^{\frac{-E_1}{kT}} + e^{\frac{-E_2}{kT}} + 4 \cdot e^{\frac{-E_3}{kT}} + e^{\frac{-E_4}{kT}} \qquad (5\text{-}12)$$

One final note: Sometimes *energy levels* are referred to as *energy states*. They are the same thing. The key is the use of the word *energy* to distinguish between all possible states of the system, including those states with the same energy, versus only the energy states (i.e., levels) of the system and using degeneracy to account for other states of the system.

In an effort to keep the distinction clear, we will usually use the word *levels* when talking about energy and the word *states* when including all configurations of the system (not just energy states, but conformations, vibrational modes, rotational modes, etc.). Sometimes, however, it will make sense for us to use the term *energy states*. Just keep in mind that *energy states* and *energy levels* mean the same thing.

The Reference State and Relative Energy

It is often convenient to define a *reference state* of the system and define all energy levels relative to the energy of the reference state. For example, when using statistical mechanics to examine the unwinding of the DNA double helix, we might define the fully twisted, closed double helix as the reference state. All other relevant states of the system then involve partially or fully unwinding the DNA double helix. Typically the reference state is chosen as the lowest energy state of the system and is therefore sometimes also called the *ground state*. The energy terms in our statistical mechanical calculations are then always the difference between the energy of a particular state and the ground state. Since these involve energy differences, we write them as ΔE_i, where

$$\Delta E_i = E_i - E_{ref} \qquad (5\text{-}13)$$

and E_{ref} is the energy of the reference state. One of the advantages of defining a reference state is that we don't have to know E_{ref}. We only need to know or measure the energy change in going from one state to the next, something that experimentally is often much simpler to do than measuring the absolute energy of a particular state.

With a reference state defined this way, we write the partition function as follows:

$$Z = 1 + \sum_{i=1}^{L_{MAX}} \omega_i e^{-\frac{\Delta E_i}{kT}} \qquad (5\text{-}14)$$

Notice the introduction of 1 as the first term in the equation. This comes from the fact that the energy difference between the reference state and itself is zero: $\Delta E_{ref} = E_{ref} - E_{ref} = 0$, and any number raised to the zeroth power is equal to 1. Notice that we are still counting from $i = 1$. We do this simply as a matter of convenience by defining the reference energy level as level zero. Thus, we can write the energy of the reference state as E_0 and $\Delta E_i = E_i - E_0$. We have also defined the reference state as having no degeneracy (again a choice of convenience), but if it did have degeneracy, we could write

$$Z = \omega_0 + \sum_{i=1}^{L_{MAX}} \omega_i e^{-\frac{\Delta E_i}{kT}} \qquad (5\text{-}15)$$

where ω_0 is the degeneracy of the reference energy state.

Gibbs Energy and the Biophysical Partition Function

When applying statistical mechanics in biophysics, we are typically interested in conformational transitions in large biomolecules (such as DNA, proteins, and lipids) or binding interactions of biological significance. In such cases the relevant energy change is the Gibbs energy change. Other energy changes, for example, those between vibrational states, are very small relatively to the Gibbs energy change of binding or of a conformational transition. Therefore the much smaller energy changes are negligible and can be ignored. Practically speaking, they contribute only to the degeneracy of a particular Gibbs energy level. Therefore it is most common in biophysical statistical mechanics to write the partition function as follows:

$$Z = 1 + \sum_{i=1}^{L_{MAX}} \omega_i e^{-\frac{\Delta G_i}{kT}} \qquad (5\text{-}16)$$

where ΔG_i represents the Gibbs energy change in going from the reference state to the ith state of the system.

QUIZ

Refer to the text in this chapter if necessary. Answers are in the back of the book.

1. Which of the following statements about statistical mechanics are true?

    S1. Statistical mechanics is the branch of mechanics that focuses on statistics.

    S2. Statistical mechanics provides a molecular interpretation of a system.

    S3. Statistical mechanics provides the mechanical tools necessary to measure statistical properties of molecules.

    S4. Statistical mechanics relates statistical averages of particles (atoms, molecules, or residues) to the overall thermodynamic measurements of a system.

    A. S1 and S2
    B. S3 and S4
    C. S1 and S3
    D. S2 and S4

2. How many ways are there to distribute 8 units of energy among five molecules? (*Hint:* Reread the section Finding Distributions.)

    A. 1
    B. 3
    C. 7
    D. 35

3. Given the distribution one molecule with 4 units of energy and four molecules with 1 unit of energy each, how many ways are there to arrange the molecules in the same distribution?

    A. 5
    B. 10
    C. 15
    D. 20

4. Given the distribution one molecule with 3 units of energy, one molecule with 2 units of energy, and three molecules with 1 unit of energy each, how many ways are there to arrange the molecules in the same distribution?

    A. 5
    B. 10
    C. 15
    D. 20

5. What is the probability of finding the system in each distribution of the three distributions, a, b, and c, shown in Fig. 5-4?

    A. 4%, 12%, and 4%
    B. 20%, 60%, and 20%
    C. 10%, 80%, and 10%
    D. .2%, .6%, and .2%

6. Given a system with a certain total amount of energy and a certain total number of molecules, the system will be *most* likely to distribute its energy among its molecules according to

   A. which distribution has the most number of permutations.
   B. which energy has the most number of distributions.
   C. which molecule has the most energy.
   D. which distribution has the most number of molecules.

7. Suppose we have a particular distribution of energy among a set of molecules. For convenience we will call it distribution X. How do we calculate the probability of distribution X relative to all possible distributions of the energy for this set of molecules?

   A. Number of permutations for distribution X divided by total number of distributions
   B. Number of distributions for distribution X divided by the total number of distributions
   C. Number of distributions for distribution X divided by the total number of permutations for all possible distributions
   D. Number of permutations for distribution X divided by the total number of permutations for all possible distributions

8. The Boltzmann distribution is

   A. the most probable distribution for any significant number of molecules.
   B. not the only distribution that occurs.
   C. so much higher in probability than all other distributions, that for all practical purposes, with any significant number of molecules it can be considered to be the only distribution that occurs.
   D. all of the above

9. A biophysicist wants to use a two-state approximation to model the conformational state of a solution of lipids. That is, we assume that the lipids can exist in only two energy states. Assuming a Boltzmann distribution and using $E_1 = 2.5 \times 10^{-20}$ J and $E_2 = 3.0 \times 10^{-20}$ J, what is the fraction of lipids in each energy level at 295 K?

   A. $F_1 = 77.3\%$ and $F_2 = 22.7\%$
   B. $F_1 = 22.3\%$ and $F_2 = 77.7\%$
   C. $F_1 = 76.3\%$ and $F_2 = 23.7\%$
   D. $F_1 = 89.0\%$ and $F_2 = 11.0\%$

10. When an energy level has multiple states with the same amount of energy, then that energy level is said to be

    A. diminished.
    B. decadent.
    C. degenerate.
    D. depraved.

chapter 6

Forces Affecting Conformation in Biological Molecules

We now turn our attention to the various forces that affect biomolecules, stabilizing or destabilizing their various conformations. Wherever there is a force, you should think of energy as well. The stronger a particular force is, the larger is the energy difference associated with a movement with or against that force. Therefore all of these forces affect the energy state of the biomolecules. It is important to understand that biomolecules run the gamut from molecules that are made up of only a few atoms, to relatively enormous *macromolecules* that are made up of literally thousands or tens of thousands of atoms. As a result, particularly in the larger molecules, it is possible that many different forces can come into play at the same time. All of the forces we will speak about are electronic in nature (i.e., they ultimately involve the interaction of positive and/or negative charges). These forces manifest themselves in a large variety of ways and we classify them as different forces in order to categorize them according to their different behavior.

CHAPTER OBJECTIVES

In this chapter, you will

- Learn the forces that influence the conformation of biomolecules.
- Learn the relative strengths of these forces.
- Understand the conditions under which each force becomes significant.
- Learn to calculate the value of various forces in biomolecules.

Chemical Bonds

Molecules are atoms connected together. The bonds that connect atoms together in molecules are called *chemical bonds*. There are, however, many types of bonds and forces that influence the exact location of the atoms within a molecule, and so influence the conformation (shape) of the molecule. These we will discuss later in this chapter. But we need to first understand chemical bonds, since chemical bonds provide the primary structure of molecules.

A chemical bond between two atoms always involves one or more of the outer electrons of the atoms. These outermost electrons that participate in chemical bonding are called *valence electrons*. There are two main types of chemical bond: A *covalent bond* occurs when one or more of the valence electrons from each atom are shared between the two atoms. An *ionic bond* occurs when one or more of the valence electrons are completely stripped away from an atom and are "donated" to the neighboring atom.

Just as a reminder (you've probably heard this before in an introductory chemistry or physics course) a couple of points to keep in mind: First, we think of electrons orbiting the nucleus as kind of an electron cloud. One possible reason for this is because, from our perspective, each electron moves so fast and orbits the nucleus in so many different directions that even in the simplest atom, hydrogen, the single electron orbiting the nucleus can be thought of as a cloud of negative charge buzzing about. From a quantum mechanical point of view, we say that the cloud represents the probability of finding the electron at a given point in space. The denser the cloud at any given point, the higher the probability of finding the electron at that point at any moment in time. From a colloquial point of view, we say the electron, in its buzzing about, "on average spends more time" where the cloud is dense and "on average spends less time" where cloud is thin.

The other point to keep in mind is that the electrons orbit the nucleus in layers called *orbitals*, with various numbers of electrons in each orbital. As we mentioned, it is the outermost electron cloud that participates in chemical bonding, with one or more of its electrons.

Covalent Bond

There is always an attractive force, in an atom, between the negatively charged electron cloud and the positively charged nucleus. In the case of a covalent bond, two atoms come close enough together (under the right conditions) for one or more outer electrons from each atom to begin orbiting both nuclei of the atoms. Electrons orbiting both nuclei are said to be shared. Covalent bonds always involve sharing pairs of electrons (one from each atom for every pair of electrons that is shared). Both nuclei feel an attractive force toward the shared electrons. Like two people wrapped in a single blanket, the force between the shared electrons and the two nuclei is the chemical bond that holds the two atoms together.

In practice, the electrons in a covalent bond may not be equally shared by both atoms. One of the atoms may have a stronger pull on the electrons than the other. This can be because one atom has a more positively charged nucleus or because its nucleus is more exposed to, or closer to, the shared electrons. When this happens, the electron cloud of the shared electrons leans closer to the atom with the stronger pull, giving that atom a slightly negative charge. The other atom will have a slightly positive charge. In such a case the covalent bond is said to be *polar*, meaning that the electric charge is not evenly distributed across the bond. See Fig. 6-1.

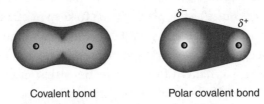

Covalent bond   Polar covalent bond

FIGURE 6-1 • A covalent bond results when one or more outer electrons (from each atom) are shared by (orbit) both atoms (left). A polar bond is a type of covalent bond where one atom has a stronger pull on the shared electrons (right).

Ionic Bond

In the extreme case of a polar bond, one of the atoms may have such a strong pull on the electrons that an electron is completely stripped away from the "weaker" atom. This "lost" electron, instead of being shared, only orbits the atom with the strong pull. The end result is that each atom becomes an *ion*. The atom that loses an electron becomes a positive ion. While the atom that gains an electron becomes a negative ion. The force holding the two atoms together no longer results from both nuclei being attracted to the shared electrons. Instead it results from the electrostatic attraction between ions, one positively charged and the other negatively charged.

Notice that it's not so black and white between covalent and ionic bonds. At one extreme, the bonding electrons are equally shared between the two atoms. In the middle case, the covalent bond becomes a polar bond. Within the realm of polar covalent bonds it is possible to have various degrees to which the electrons lean more toward one atom than the other, creating various degrees of polarity in the bond. In the extreme case one or more electrons are no longer shared at all. You can see that there is somewhat of a continuum between completely evenly shared electrons (covalent bonds) and completely ionized atoms (ionic bonds). In practice, however, if the electrons are shared even slightly, then the bond is referred to as covalent (although it may be highly polar). Only when an electron is completely transferred from one atom to another do we refer to the bond as ionic.

Ionic bonds are most common in crystalline solids. But in fluid solutions, such as inside cells, ions typically float about somewhat freely. Each ion experiences random attractive and repulsive forces from neighboring ions that are also floating about. This is due to the thermal (kinetic) energy of the molecules.

There are cases where ions in solution bond to specific ionized regions of biomolecules. One example is the case of sodium (Na^+) ions binding to the negative phosphate groups (PO_4^-) on the backbone of the DNA double helix. The sodium ions help to stabilize the double-helical structure by reducing the repulsive force between the negatively charged phosphate groups along the DNA backbone. This stabilization effect can be demonstrated experimentally by measuring the amount of energy it takes to unwind the DNA helix at various salt (sodium) concentrations. As the sodium concentration is increased, it takes more energy to unwind the DNA helix. The fact that more energy is needed to disrupt the DNA helix indicates that the helix is more stable at higher sodium concentrations. This implies that somehow sodium stabilizes the helical structure.

We will explore molecular conformations and their stability in detail in the next chapter. The case of sodium ions and the DNA backbone was brought here only as an example of ionic behavior in solution. Ions play an important role in many biological processes and affect the stability of molecular conformations. However, the binding of ions in solution is more aptly treated as a case of ligand binding than as ionic bonds in crystalline solids. Therefore, we will not concern ourselves with ionic bonds as such when discussing biological systems. Rather, going forward, when we refer to chemical bonds, it should be understood that we mean covalent bonds.

Movement in Chemical Bonds and Molecular Conformation

Covalent bonds typically have various amounts of flexibility. This flexibility, depending on the particular bond, can allow for movements such as free rotation (like a wheel on an axle), and springlike movements such as bending, compression, extension, and twisting. See Fig. 6-2.

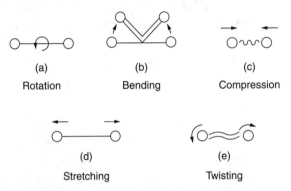

(a) Rotation
(b) Bending
(c) Compression
(d) Stretching
(e) Twisting

FIGURE 6-2 • Types of movements commonly found in chemical bonds. (a) Free rotation. (b) Bending. (c) Compression. (d) Extension or stretching. (e) Twisting (limited rotation). Motions (b) through (e) are all springlike motions: They occur with a restoring force, a force that somewhat opposes their motion and tends to restore the atoms back to their original position.

The springlike motions (bending, compression, extension, and twisting) occur with a restoring force. That is, just like a spring, the chemical bond exerts a force that somewhat opposes their motion and tends to restore the atoms back to their original position. When a force causes one of these motions in a molecule, the balance between the force causing the motion and the restoration force can also result in vibration of the atoms back and forth. This can give rise to various vibrational states in the molecule, where particular covalent bonds vibrate at specific frequencies.

All of these motions, either alone or together, allow for a large number of conformations even in relatively simple biomolecules. For example, imagine a simple molecule consisting of four atoms connected together, one after the other. Now let's say these particular covalent bonds allow free rotation and bending. Figure 6-3 shows some possible conformations for this molecule. To help with our discussion, the atoms in Fig. 6-3 are labeled 1, 2, 3, and 4, and the different conformations are labeled A, B, C, and D.

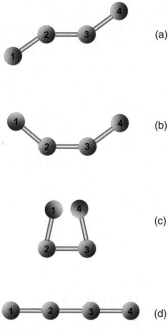

FIGURE 6-3 • Some possible conformations of an imaginary molecule made of four atoms connected together in a line. The atoms are labeled 1, 2, 3, and 4. The different conformations are labeled A, B, C, and D.

If there is a strong attraction between the two outside atoms (atoms 1 and 4), then conformation C might be the most favorable: The strong attraction draws atoms 1 and 4 as close together as they can get without breaking any of the covalent bonds.

If the attraction between atoms 1 and 4 is strong but the bonds are somewhat rigid, allowing only a limited amount of bending, then a conformation similar to B would be more likely: Atoms 1 and 4 are drawn together, but only as far as the partially flexible bonds will allow. If on the other hand there is a repulsive force between atoms 1 and 4, then we might see a conformation such as A or D, where, as a result of the repulsive force, atoms 1 and 4 prefer to be as far away from each other as possible.

Notice we haven't said anything yet about what forces might account for such an attraction or repulsion between the different atoms of the molecule. That's what the rest of this chapter is about. The important point here is to get a concept of how various bond movements, combined with forces, can result in a number of different conformations for even the simplest of molecules. You can imagine how many different conformations might be possible in a molecule made of hundreds or even thousands of atoms.

Bond movements can also allow for multiple paths between two different conformations. Look for example at conformations A and B in Fig. 6-3. There are two different movements that can take the molecule from conformation A to conformation B. One possibility is free rotation of the bond between atoms 2 and 3 (see Fig. 6-4a). To get from conformation A to conformation B, let the bond angles stay constant (no bending). Hold atoms 3 and 4 in place, and turn atom 2 like a wheel with the bond between atoms 2 and 3 acting as an axle. Atom 1 will rotate up until the molecule is in conformation B.

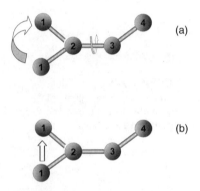

FIGURE 6-4 • Two ways to go from conformation A to conformation B. (a) Rotation. (b) Bending.

Another possibility is bending (see Fig. 6-4b). To get from conformation A to conformation B through bending, don't allow the bonds to rotate. Instead straighten out the bond angle between atoms 1, 2, and 3, and then bend it upward in the opposite direction. The result is converting conformation A into conformation B.

It is possible that such a molecule may allow only one of these two movements in going through the transition from conformation A to conformation B, depending on the flexibility of the specific covalent bonds. It is also possible that such a molecule may allow both motions, in which case the path that the molecule takes will depend on the Gibbs energy of the intermediate steps.

Forces That Affect Conformation in Biomolecules

All of the forces that affect conformation in biomolecules are electrical in nature, meaning they result from attractions or repulsions between charges. However, the specifics of how these attractions or repulsions manifest themselves can vary quite a bit depending on the situation. It is convenient and helpful to classify the various situations and give each a name as if they are actually separate forces. Just keep in mind that each "force" that we are about to describe is just a particular case or manifestation of the electromagnetic force generated by protons and electrons in the atoms making up biological molecules.

We have already described covalent and ionic bonds. Covalent bonds play a major role in determining the primary structure of biomolecules. In Chap. 9 we will see covalent bonds also contribute to secondary and tertiary structure of some molecules via *cross-linking*, which is the covalent connecting of otherwise distance parts of a molecule.

Ionic bonds, as we mentioned, specifically refer to the arrays of bonds between ions in a crystalline solid. Generally these will not concern us since the natural state of biomolecules is not a crystal. (One exception where we might be interested in ionic bonds in biomolecules is in the context of the biophysical technique of X-ray diffraction which requires crystallized samples of biomolecules.) On the other hand there are two cases where we are very interested in the forces between ions because they play a significant role in influencing the conformation of biomolecules. One is the attraction and repulsion of ions in solution (both individual ions and ionized portions of biomolecules) which we mentioned is most aptly treated as a case of ligand binding. The other is the formation of salt bridges, which can be thought of as single ionic bonds (as opposed to the arrays of ionic bonds found in crystals) that hold together otherwise distant parts of a biomolecule.

All of the forces we are about to describe exert themselves both between molecules and between different parts of the same molecule. The general term for molecular forces other than covalent and ionic bonds is *van der Waals forces,* named for Dutch physicist Johannes van der Waals. You should be aware that textbooks vary somewhat in how they use the term *van der Waals force.* Some use the term very generally as meaning all noncovalent intermolecular forces, including ionic bonds. Others may limit its use to a subset of the forces described in the following sections. We will consider *van der Waals forces* to include all molecular forces that are not covalent and not crystalline ionic bonds.

In this chapter and the next, our focus is the influence of these forces on the conformation of biomolecules. This influence on shape arises primarily from the forces as they are within the molecule, that is, between different parts of the molecule whose conformation we are considering. Forces between separate molecules can also play a role in molecular shape but, with some exceptions, tend to be less significant with regard to conformation than those within the molecule itself.

On the other hand these same forces, when between separate molecules, are important for molecular associations such as ligand binding and quaternary structure. In this case it is the forces *between* molecules that play the largest role, whereas forces within the molecule contribute to molecular associations indirectly by their influence on conformation.

Charge-Charge Forces

The force between two electrostatic charges is defined by Coulomb's law.

$$F = k_e \frac{q_1 q_2}{r^2} \tag{6-1}$$

where F is the force, q_1 and q_2 are two charges, r is the distance between the two charges, and k_e is a constant. In SI units, k_e is equal to 8.98755×10^9 Nm^2/ C^2 (newton meter^2 per coulomb^2). The constant k_e is also equal to $1/(4\pi\varepsilon_0)$ where ε_0 is a constant known as the *permittivity of a vacuum*.

In biophysics, we are most often interested in the amount of energy, or potential energy, resulting from molecular forces. This is because potential energy resulting from a force is that force's contribution to the Gibbs energy of the molecule. And it is the Gibbs energy that drives biophysical processes. When a molecule has a choice between two possible configurations, then (as you learned in the Chap. 4), the configuration that decreases or minimizes the Gibbs energy will be favored. If we can calculate the potential energy resulting from the various forces that influence molecular conformation, then we can determine the most likely conformation as the one that most reduces the Gibbs energy.

Let's write Coulomb's law in a form that defines the potential energy due to the electrostatic interaction. Since potential energy is the potential to do work and work is force times distance, we can convert Eq. (6-1) to its potential energy form by multiplying both sides by the distance between the two charges.

$$F r = k_e \frac{q_1 q_2}{r^2} r \qquad (6\text{-}2a)$$

On the left side we now have force times distance, which is work and which is also equal to the amount of potential energy to do that work. (Whether we treat force times distance as work or as potential energy depends only on our perspective. For example, pushing a ball up a hill takes work, but the potential energy of the ball has increased. Also, strictly speaking, for those who know calculus, we really should integrate both sides of the equation over the distance, since the force is changing with distance. However, mathematically in this case the formula we would end up with would be the same, and conceptually multiplying by distance is simpler.)

Now we can simplify the equation, by replacing force times distance with potential energy on the left and canceling out the extra distance term on the right.

$$U = \frac{q_1 q_2}{\varepsilon r} \qquad (6\text{-}2b)$$

where U is the potential energy, and we have replaced the constant k_e with $1/\varepsilon$ which includes $4\pi k_e$ and is a property of the medium in which the charges are placed. The value ε is called *permittivity* and is a measure of how easily a medium is polarized by an electric field. ε is also called the *dielectric constant* of the medium.

It is often convenient to express the *dielectric constant* or *permittivity* of a medium relative to the permittivity of a vacuum. In other words, we define the permittivity of a vacuum as having a value of exactly 1 and the permittivity of all other substances as calculated relative to that of a vacuum. For example, the permittivity of air is 1.000585 and the permittivity of paper is about 3. The permittivity of lipids that make up the cell membrane is about 2. And the permittivity of water is about 80. (Note that permittivity is temperature dependent, and these values are for 20°C which is about room temperature.)

Notice that the dielectric constant of water is very high. This reflects water's exceptional ability to conduct electricity. Notice also from Eq. (6-2b) that increasing the permittivity, ε, will decrease the potential energy U. In other words, a high dielectric constant reduces the potential energy and reduces the force between two charges. This is because a medium that is easily polarized by a charge placed within it will orient its own polarity so as to oppose the force due to the charge. For example, if we place a positive charge into a medium of neutral but polar molecules, those molecules will orient themselves with their negative ends toward the positive charge. If we also place a negative charge into the same medium, the orientation of the polar molecules will effectively reduce the attraction (force and potential energy) between the positive and negative charge in the medium. See Fig. 6-5.

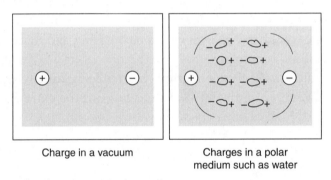

Charge in a vacuum   Charges in a polar
                                       medium such as water

FIGURE 6-5 · A positive and negative charge in a vacuum and in a medium of high permittivity. The neutral, but polar, molecules of the high-permittivity medium (e.g., water) orient themselves with their negative ends facing the positive charge and their positive ends facing the negative charge. This orientation of the polar molecules creates small electric fields that oppose the electric fields of the positive and negative charge in the medium. This effectively reduces the force, and potential energy, between the positive and negative charge.

The important consequences of permittivity in for biological systems cannot be understated. Ions, ionized portions of molecules, and highly polar molecules will prefer to be in regions of high permittivity where the high dielectric constant reduces the Gibbs energy. On the other hand, regions of low permittivity, such as cell membranes, will tend to exclude ions due to the cost of increasing the Gibbs energy to place ions there. Ion channels (certain proteins within cell membranes) allow ions to pass through membranes by creating a tunnel of high permittivity through the membrane.

Dipole Forces

An *electric dipole* is defined as two equal and opposite charges separated by a fixed distance. In many neutral molecules, charge is not evenly distributed within the molecule. This can be due to the uneven distribution of electron clouds, as we mentioned when we discussed polar covalent bonds. It can also be due to different parts of a molecule being ionized differently. Either way, the end result is that there are many situations in biophysics where the behavior of a molecule or a portion of a molecule is very easily approximated by treating it as an electric dipole.

Just as amount of charge is a convenient way to characterize an ion, a convenient way to characterize a dipole is by a quantity called the *dipole moment*. The *dipole moment* is equal to the magnitude of the charge at one end of the dipole, times the fixed distance between the two charges. Let q_d be the magnitude of the charge at one end of the dipole, and let ℓ be the distance between the two charges (i.e., the length of the dipole), then the dipole moment is given by $d = q_d\ell$.

We will examine several interactions involving dipole forces.

- Charge-dipole forces
- Dipole-dipole forces
- Induced dipoles
  - Charge-induced dipole forces
  - Dipole-induced dipole forces

Charge-Dipole Forces

Suppose we have a neutral but polar molecule (or portion of a molecule). We can represent this molecule as a dipole of two opposite charges $q_d{}^+$ and $q_d{}^-$, separated by a distance ℓ. See Fig. 6-6.

The potential energy between an ion, or point charge, q_{ion} and the dipole, is simply the sum of the potential energy between the ion and each of the charges $q_d{}^+$ and $q_d{}^-$ in the dipole.

$$U = U_{ion,q_d+} + U_{ion,q_d-} = \frac{q_{ion}q_d{}^+}{\varepsilon r_{d^+}} + \frac{q_{ion}q_d{}^-}{\varepsilon r_{d^-}} \tag{6-3}$$

where $r_d{}^+$ is the distance between the ion and the positive end of the dipole, and $r_d{}^-$ is the distance between the ion and the negative end of the dipole.

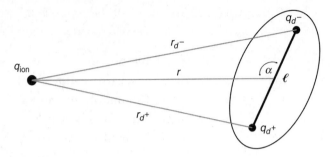

FIGURE 6-6 • Interaction between a charge q_{ion} and a molecule with a dipole moment $d = q_d\ell$.

The whole point of defining a dipole is to be able to treat the dipole as a single entity, instead of as two separate charges. This allows us to simplify the expression for the potential energy. Instead of two separate distances r_d^+ and r_d^-, we introduce only the distance between the ion and the center of the dipole r. And, since the charges on the dipole are equal and opposite, we replace q_d^+ with q_d and q_d^- with $-q_d$.

We can approximate the distance r in terms of r_d^+ and r_d^- by the projections of r_d^+ and r_d^- onto r. In other words, $r_d^+ = r + (1/2)\ell \cos \alpha$, where ℓ is the length of the dipole and α is the angle of orientation between dipole axis and a line from the ion to the center of the dipole, as shown in Fig. 6-6. This approximation holds true whenever ℓ is significantly less than r.

We can now write the potential energy between a charge and a dipole as

$$U = \frac{q_{ion}q_d}{\varepsilon[r+(1/2\ell)\cos\alpha]} + \frac{q_{ion}q_d}{\varepsilon[r-(1/2\ell)\cos\alpha]} \tag{6-4a}$$

or

$$U = \frac{-q_{ion}d\cos\alpha}{\varepsilon[r^2 - (\ell^2/4)\cos^2\alpha]} \tag{6-4b}$$

where $d = q_d\ell$ is the dipole moment. Now we have the potential energy in terms of the charge on the ion, the dipole moment, and the distance between the ion and the dipole.

We can simplify this equation even further in the case where ℓ is much less than r. If ℓ is much less than r, then $(\ell^2/4)\cos^2\alpha$ is much less than r^2, so the $(\ell^2/4)\cos^2\alpha$ term can be ignored. This gives us

$$U = \frac{-q\,d\cos\alpha}{\varepsilon r^2} \tag{6-5}$$

The approximations introduced into Eq. (6-5) hold true in many biophysical problems. The most common situations where we treat charges as dipoles involve two atoms connected by a polar covalent bond. The dipole then interacts either with an ion in solution or with an ion on some relatively distant part of the molecule that has folded back to come closer to the dipole. The distance, however, between adjacent covalently bonded atoms is really quite small, much smaller than between these atoms and an ion in solution and also much smaller than the distance between these atoms and some other part of the molecule that has folded back to interact with the dipole. Therefore it is typically the case that ℓ is much less than r.

Equation (6-5) makes it easy to see the most important points to remember about charge-dipole forces. These points still apply even in the case where ℓ is not much less than r.

- Charge-dipole potential energy depends on the orientation of the dipole relative to the charge (i.e., the angle α in Fig. 6-6).
- The charge-dipole potential energy is inversely proportional to the *square* of the distance between the charge and the dipole.

Notice that the charge-dipole potential energy has a higher inverse power relationship with distance $(1/r^2)$ than does the coulomb potential $(1/r)$ between two charges [Eq. (6-2b)]. This means that, in the case of a charge-dipole interaction, the magnitude of the potential energy decreases much faster with distance than it does in the case of a charge-charge interaction. In other words charge-dipole interactions have a much shorter range than charge-charge interactions.

This illustrates a general trend that, when multiple charges interact simultaneously, the more charges there are, the higher the power on the inverse relationship with distance and the shorter the range of interaction.

Thermal Averaging

Equation (6-5) assumes a fixed orientation of the dipole with the charge; the angle α is assumed constant. This is a safe assumption in many situations, for example, in the case of a relatively inflexible folded protein molecule, where a dipole on one part of the molecule interacts with a charge on another part of the molecule.

But in a situation where the dipole is on one molecule and the charge on another, and the molecules are in solution, then the dipole can freely rotate in solution. In such a case we have to use the average potential energy, averaged over all possible orientations.

In calculating the average, we weight each orientation by the probability of finding the dipole in that particular orientation. Statistical mechanics tells us this probability is given by the Boltzmann distribution. So, analogous to Eq. (5-8) for average energy, we use the following for the average charge-dipole potential energy:

$$\langle U \rangle = \sum U_\alpha \frac{e^{-U_\alpha/kT}}{Z} \tag{6-6}$$

where the sum is over all possible orientations or values of the angle α.

If we assume that the dipole moment times the potential energy is much less than kT, then the above sum can be calculated by

$$\langle U \rangle = \frac{-q^2 d^2}{\varepsilon kT r^4} \tag{6-7}$$

A few points to note about thermal averaging:

- We call it *thermal averaging* because it accounts for the random motions of molecules in solution due to thermal energy.

- The average potential energy has an inverse relationship to temperature: As we increase the temperature of the solution we reduce the magnitude of the electrostatic potential energy. This is because we force, by thermal motion, the dipole to take on more and more random orientations even if those orientations are unfavorable from an electrostatic point of view.

- The average charge-dipole potential energy is inversely proportional to the *fourth power* of the distance between the charge and the dipole. You can see that allowing the dipole to freely rotate significantly reduces the range of interaction compared to the chare-dipole with a fixed orientation.

- The charge and the dipole moment are now squared, so although the range of interaction falls off much quicker with distance, within close range the interaction can be somewhat stronger (compared with a charge-charge interaction). This, however, is balanced by the temperature dependence which makes the interaction weaker except at very close distances.

Dipole-Dipole Forces

Without going into detail we will briefly discuss and present the formulas for dipole-dipole interactions. The main differences between charge-dipole interactions and dipole-dipole interactions are

Still Struggling

The detailed derivation of the various force equations is not necessary in an introductory biophysics textbook. One might legitimately ask then, why show the equations at all? The answer is that we do so to represent the fundamental concepts that the equations clearly illustrate. For example, the steps in going from Eq. (6-6) to Eq. (6-7) are not trivial, but you don't need to know these steps in order to understand conceptually how Eq. (6-7) was arrived at, how to use the equation, and what it tells us about thermally averaged charge-dipole interactions. From a demystified perspective, it is important to understand why we do thermal averaging in the first place (to account for the random motion of molecules and how that motion affects the energy associated with the particular force interaction). And it is important to know that someone took the time to derive the result, and to know what we learn from that result; namely that accounting for the thermal motion of molecules tells us that the interaction falls off more rapidly with distance, but is stronger at closer distances. Throughout this chapter we emphasize the important points to note and learn from each of the equations. You should focus on learning these points and on understanding generally how each equation was arrived at and in which situations each equation is important.

- When two dipoles interact, we have a total of four charges interacting, as compared with three charges in the case of a point charge and a dipole.
- When two dipoles interact, there are two different angles to consider for *each* dipole, as compared with a single angle to consider in the orientation between a point charge and a dipole.

The two angles to consider are

1. Each dipole can orient itself toward or away from the other dipole (as we also saw in Fig. 6-6 with a charge and a dipole).
2. Each dipole can turn to the right or left (clockwise or counterclockwise) relative to the other dipole.

In the case of a point charge and a dipole, there was no turning to the right or left. There was only the angle toward or away from the point charge. This is because electrostatic interaction depends only on the distance from charge to charge, and turning a dipole clockwise or counterclockwise relative to a point charge does not change the distance from the point charge to each end of the dipole; only turning the dipole on an angle toward or away from the point charge changes this distance.

However, turning a dipole clockwise or counterclockwise relative to another dipole will change the distances from each end of one dipole to each end of the other dipole. So dipole-dipole interactions must consider two different angles for each dipole, making a total of four angles. See Fig. 6-7.

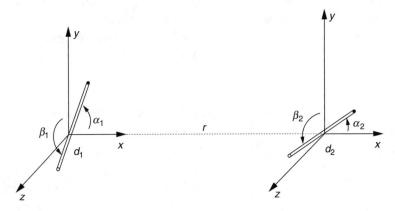

FIGURE 6-7 • Dipole-dipole interactions. Two angles must be considered for each dipole: one angle in the *xy* plane between the *x*-axis and the dipole, and the other angle in the *yz* plane between the *y*-axis and the dipole. To keep things simple, we orient the axes so that the distance *r* between the two dipoles is along the *x*-axis.

The potential energy between two dipoles can be written as

$$U = \frac{-d_1 d_2 F(\alpha_1, \beta_1, \alpha_2, \beta_2)}{\varepsilon r^3} \tag{6-8}$$

Compare this equation with Eq. (6-5). To emphasize the form of the potential energy function, we write simply $F(\alpha_1, \beta_1, \alpha_2, \beta_2)$, where F is some function of the four angles $\alpha_1, \beta_1, \alpha_2, \beta_2$ (see Fig. 6-7). The important points to remember here are

- The potential energy falls off with the *cube* of the distance r^3 as compared with the square of the distance for a charge-dipole interaction, and compared with $1/r$ for a charge-charge interaction. This demonstrates the trend mentioned earlier that when multiple charges interact simultaneously, the more charges there are, the higher the power on the inverse relationship with distance, and the shorter the range of interaction.

- For a fixed orientation between dipoles (such as one may find in a folded protein or molecular complex), the potential energy depends on this orientation, and the orientation is defined by four different angles.

Thermal Averaging in Dipole-Dipole Interactions

When the orientation between two dipoles is not constant, but is allowed to move freely by the random thermal motion of molecules, then thermal averaging can be done using the Boltzmann distribution to calculate the average over all possible orientations, as was noted for the charge-dipole interaction. For the case of dipole-dipole interactions, the average potential energy is given by

$$\langle U \rangle = \frac{-2}{3} \frac{d_1^2 d_2^2}{\varepsilon kT r^6} \tag{6-9}$$

Notice that (as was the case with a charge-dipole interaction) thermal averaging has doubled the power on the inverse distance dependence compared with the fixed orientation case. Dipole-dipole potential energy can indeed be significant, but only at very short distances.

Induced Dipoles

The electron cloud of a neutral, nonpolar molecule can be distorted in the presence of an electric field so that the molecule becomes polar. This is called an *induced dipole*. The electric field that induces the polar quality of the molecule typically comes from a nearby ion or dipole.

We can measure the tendency of the electron cloud of an atom or molecule to be distorted from its normal (nonpolar) shape by an electric field. We call this measurement *polarizability*, defined as the ratio of the induced dipole moment to the electric field strength.

$$\alpha_p = \frac{d}{E} \tag{6-10}$$

Notice that what this equation says is that a large degree of *polarizability* means a given electric field induces a large dipole moment. A small degree of *polarizability* means the same given electric field induces a small dipole moment.

(Notice also that α_p is the symbol for *polarizability*, not to be confused with our use of α as the angle between a dipole and a point charge.)

The standard formula for the strength of an electric field due to a point charge is given by

$$E = \frac{q}{\varepsilon r^2} \tag{6-11}$$

where E is the electric field strength, q is the amount of charge, ε is the dielectric constant of the medium (also called permittivity), and r is the distance from the point charge.

If we substitute Eq. (6-11) for E in Eq. (6-10) and then solve for the dipole moment, we get

$$d = \alpha_p E = \frac{\alpha_p q}{\varepsilon r^2} \tag{6-12}$$

This is the formula for the dipole moment of an induced dipole. We can substitute this formula into Eq. (6-5), to get the potential energy due to the interaction of a charge with an induced dipole.

$$U = \frac{-\alpha_p q^2}{\varepsilon r^4} \tag{6-13}$$

In Eq. (6-13) we've made an additional assumption. We know that an induced dipole will orient itself in the direction of the inducing electric field. If we assume that the interacting point charge is also the source of the inducing electric field, then we can assume the dipole will orient itself in the direction of the charge. Making this assumption leads to $\cos \alpha = 1$ [compare Eq. (6-13) with Eq. (6-5)].

Hydrogen Bonds and Water

Water is an amazing molecule, and life as we know it could not exist without water. Entire volumes have been written about the chemical and physical properties of water. What concerns us here is that water is a highly polar molecule. The molecule is shaped somewhat like the letter V with a slight positive charge at each end and a somewhat negative charge in the middle. See Fig. 6-8.

Water is a permanent dipole. Actually, it can be treated as two dipoles, or as a V-shaped "tri-pole" with a negative charge in the middle and positive charges at the ends. Sometimes we treat each "leg" of a water molecule as a dipole,

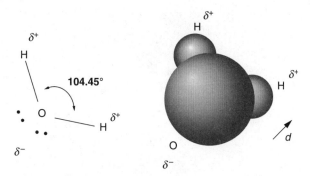

FIGURE 6-8 · A water molecule consists of an oxygen atom with two hydrogen atoms. The oxygen atom is significantly larger than the hydrogen atoms. The oxygen atom has a stronger pull on the electron cloud, resulting in a slight positive charge δ^+ on each of the hydrogen atoms and a somewhat negative charge δ^- on the oxygen atom. The direction of the dipole moment d is shown.

because typically only one side (at a time) of the water molecule interacts with other charges or dipoles. However, it is possible, for example, depending on the size and shape of the interacting molecules, for all three charges of the water molecule to interact with several charges on another molecule. In such a case we can also approximate a water molecule as a single dipole with a dipole moment, d, as indicated in Fig. 6-8. This makes sense since the sum of the partially positive charges on the hydrogen atoms is equal to the partially negative charge on the oxygen atom.

The permanent dipole nature of water causes water molecules to attract one another and to line up in or orderly arrangement. The slightly positive hydrogen atom of one water molecule orients in the direction of the negative oxygen atom of a neighboring water molecule. See Fig. 6-9. The attractive force between the positively charged hydrogen atom on one water molecule and the negatively charged oxygen atom on the other molecule is called a *hydrogen bond*. These bonds contain potential energy. When oriented as such, since the dipoles tend to orient toward one another, the interaction is effectively a charge-charge interaction. That is, because the dipoles are lined up end to end, for all intents and purposes it is as if two point charges are interacting: the slightly positively charged hydrogen on one water molecule with the somewhat negatively charged oxygen on another water molecule. Due to the orientation of the dipoles, each charge can "see" only one end of the dipole on the other molecule.

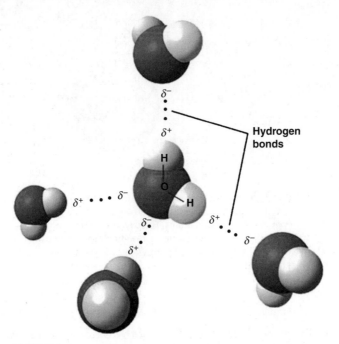

FIGURE 6-9 · Hydrogen bonding in water. (*Courtesy of Wikimedia Commons.*)

Hydrogen Bonds in General

Water is not the only molecule capable of hydrogen bonding. Whenever a hydrogen atom in a molecule has a slightly positive charge, it can interact with a negative or partially negative charge elsewhere (either on another molecule or on other parts of the same molecule).

Notice that a hydrogen bond is *not* an interaction with a single hydrogen atom or even a hydrogen molecule. A single hydrogen atom is nonpolar. A hydrogen molecule is also nonpolar, consisting of two hydrogen atoms covalently bonded together in a nonpolar bond. Even if we had a single hydrogen atom in the form of a positively charged ion, then the interaction would be ionic. This is not what we mean by hydrogen bond.

When we speak of a *hydrogen bond*, it is important to remember that the hydrogen atom is always covalently bonded to an atom other than hydrogen, typically a much larger atom. The larger atom has a stronger pull on the electron cloud, leaving the hydrogen with a slightly or partially positive charge. It is this polar covalently bonded hydrogen that can participate in a hydrogen bond by being attracted to a negative or partially negative charge.

The physical property that describes the amount of pull a particular atom has on electrons is called *electonegativity*. Atoms that have a strong pull on electron clouds are said to be highly *electronegative*. In biomolecules, the most commonly found highly electronegative atoms include nitrogen and oxygen. There are many places in proteins, DNA, and other biomolecules where hydrogen is covalently bonded to oxygen or nitrogen in a highly polar bond. This leaves the hydrogen with a partially positive charge and able to participate in hydrogen bonds.

Hydrogen bonds in biomolecules can form between different parts of the same molecule, for example, between different parts of a folded protein. They also form between two different molecules, for example, when a protein binds to DNA. And they form between biomolecules and the water that surrounds them. As we will see, all types of hydrogen bond interactions play an important role in the conformation and functioning of biomolecules, including the hydrogen bonds between water and biomolecules, and even the hydrogen bonds among the water molecules themselves that surround a biomolecule.

Aromatic Ring Structures: Cation-Pi Interactions and Stacking

In many molecules, some atoms are covalently bonded together to form a ring. The designation *ring* does not necessarily mean it is geometrically a circle. By *ring*, we simply mean that if we trace from atom to atom as each atom is covalently bonded to the next, eventually we get back to the first atom. The *ring* of atoms can take on various shapes, for example a pentagon, or hexagon, or even a kind of zig-zag shape like the top edge of a crown.

In biomolecules there is a common type of ring structure known as an *aromatic* ring. Aromatic ring structures have certain defining characteristics.

> The atoms making up the ring lie within a plane (so the zig-zag ring is *not* considered aromatic).

> All of the ring atoms contribute electrons from their *pi orbital* to the covalent bonds that hold the ring together.

It's not important for this discussion to know the details of what a pi orbital is; you only need know that an *orbital* is a way of mathematically defining a portion of the overall electron cloud. And a *pi* orbital is just the name of one particular orbital that has certain mathematical properties. The covalent bonds that hold the ring atoms together in an aromatic ring consist of electrons from both sigma orbitals and pi orbitals, but a defining

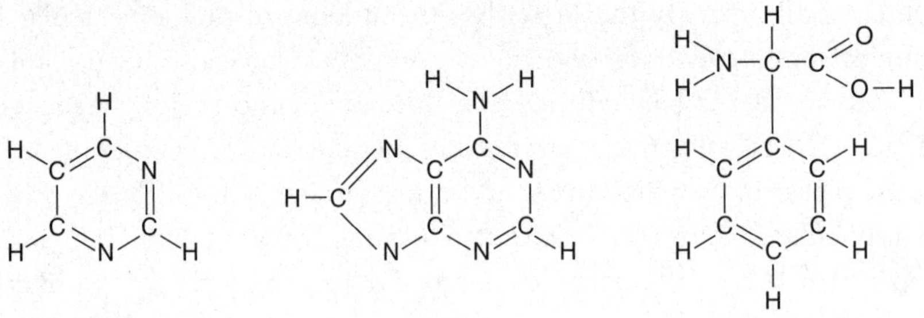

FIGURE 6-10 · Example aromatic molecules found in biological systems.

characteristic of aromatic rings is the fact that every atom in the ring contributes at least one pi electron.

In biomolecules, aromatic rings are almost always made of six atoms or sometimes five if in an ionic state. Typically most or all of the ring atoms are carbon. Sometimes one or two of the ring atoms are nitrogen, or occasionally oxygen, while the remaining atoms are all carbon. See Fig. 6-10.

Cation-Pi Interactions

The electron clouds of aromatic structures have two unique features that affect the way they interact with other molecules. First, you learned at the beginning of this chapter that two atoms covalently bonded together share electrons; that is, one or more electrons orbit both atoms in the covalent bond. Electrons from the sigma orbitals follow this rule and are shared between any two adjacent atoms in the ring. However, the electrons from the overlapping pi orbitals become *delocalized*; that is, they form a cloud around the *entire* aromatic ring and are not associated with only one or a pair of atoms on the ring. Furthermore the pi electrons are distributed mainly above and below the plane of the ring, with the sigma orbitals localized to the space around pairs of adjacent ring atoms. The result is an uneven distribution of charge, with negative charge above and below the plane of the ring and positive charge along the edges. See Fig. 6-11.

The electronegative layers above and below the ring plane interact strongly with cations. Furthermore, depending on the polarizability of the aromatic ring, a cation may accentuate the uneven charge distribution. The layers of positive and negative charge are easily modeled as two dipoles as shown in Fig. 6-11.

Interactions between cations and aromatic rings are common in proteins that contain aromatic structures. Such *cation-pi interactions* are found in the binding

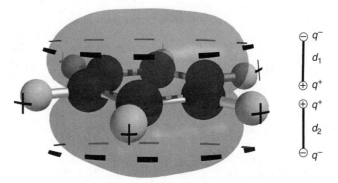

FIGURE 6-11 • In aromatic rings, the electron cloud is distributed so that a layer of partially negative charge exists above and below the plane of the ring, with partially positive charge along the ring edge. The electronegative layers above and below the ring plane can interact strongly with cations (positive ions).

of ligands to proteins and in the binding of proteins to one another, as well as within protein conformations where an aromatic portion of a protein folds over and interacts with a cation side chain on the same protein. Mathematics similar to the dipole interactions outlined previously show that the energy of cation-pi interactions decreases in proportion to the cube of the distance. That is, the energy of cation-pi interactions is proportional to $1/r^3$.

Stacking

The distribution of pi electrons in aromatic rings is not static, but fluctuates in time. These fluctuations momentarily create regions of positive and negative charge on the plane surface of the ring. The cause may be due to the movement of a passing ion or may be induced by other nearby fluctuating electron clouds. When the surfaces of two aromatic rings approach each other, fluctuations in the pi electron cloud of one ring can induce reciprocal fluctuations in the other. This creates multiple regions of alternating partially positive and partially negative charge on the surface of the ring. The electron cloud shifts back and forth among these regions in harmonic motion, like a spring or seesaw. See Fig. 6-12. When the aromatic rings are close enough together, the fluctuations in each ring induce and influence fluctuations in the other so that it is possible for the charge movements to come in sync with each other.

Stacking interactions occur when two aromatic rings are stacked on top of each other, like plates, and the reciprocal alternating fluctuations in the electron clouds create an attractive force between adjacent rings. Where one electron

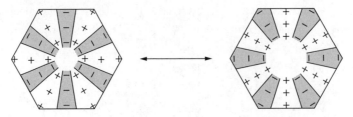

FIGURE 6-12 • Electron cloud fluctuations in aromatic rings cause alternating regions of partially positive and partially negative charge. When two aromatic rings are stacked like plates, the opposite alternating fluctuation in adjacent rings creates an attractive force between the rings.

cloud is thin and the ring surface exhibits a partially positive charge, at the same time in the same place on the adjacent ring, the electron cloud is thick and exhibits a partially negative charge. The partially positive and partially negative charges attract one another. A moment later the positive regions fluctuate to negative, and vice-versa, so the stacked rings continue to attract each other. This is because the rate, or frequency, of the fluctuations is the same in adjacent rings, since fluctuations in each ring induce and accentuate the opposite fluctuations in the other ring.

Aromatic stacking interactions are common in nucleic acid structures such as DNA. They account for the majority of the force that holds DNA into its helical shape.

Dispersion Forces

Aromatic rings are not the only molecular structures that exhibit fluctuations in their charge distribution. In fact all electron clouds fluctuate rapidly, and in many molecules this leads to momentary dipole moments. When two such molecules are near each other, the fluctuating dipole moments can lead to attractive or repulsive forces. In order for these forces to be significant (to last more than the momentary fluctuation), the fluctuations of the two molecules need to be in sync with one another. That is, the rate or frequency of fluctuations needs to be the same in both molecules. This is possible if the structures are somewhat similar and if their polarizability allows fluctuations in one structure to induce fluctuations in the other. This is exactly the case, as mentioned, with stacking interactions. The charge distribution fluctuations in one aromatic structure induce synchronous fluctuations in the adjacent aromatic structure.

Forces that arise from in-sync fluctuating charge distributions are called *dispersion forces*. There are various approaches to modeling fluctuating dipole moments and the mathematics can get rather complicated. In general they are treated somewhat similarly to freely rotating dipoles, with an added complication that electron cloud fluctuations are much faster than molecular rotation. The energy associated with dispersion forces varies as the 6th power of the reciprocal of the distance, $1/r^6$, so it is a relatively short-distance interaction.

Steric Interactions

Steric interactions or *steric repulsions* are simply terms that describe atoms or molecules bumping into each other. It's another way of saying that two things can't be in the same place at the same time. Physically what's happening is that the electron cloud of one atom begins to penetrate the electron cloud of another. This leads to a very sharp rise in free energy, making it very unlikely that the penetration will continue any further. The energy of steric repulsions varies as $1/r^{12}$.

The potential energy, or free-energy function, of steric interactions combined with dispersion forces looks typically like the one shown in Fig. 6-13. As the molecular structures approach each other, at a very close distance, dispersion forces create an attraction and the potential energy drops. Then as the molecules move even closer together, the electron clouds begin to penetrate each other and the potential energy rises sharply. This is the *steric interaction*. The potential energy

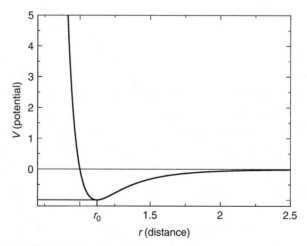

FIGURE 6-13 · Potential energy as a function of atomic distance. As two molecules approach each other, dispersion forces create an attraction. At a distance of r_0, the potential energy reaches a minimum. At this distance the dispersion forces are at their strongest. At distances closer than r_0, *steric interactions* begin to take over; the electron clouds begin to penetrate one another and the potential energy rises sharply, making it increasingly difficult for the molecules to get any closer together.

quickly becomes so large that the molecules do not have enough energy to move any closer together. The steric interaction (repulsion) keeps them apart.

This illustrates the general principle that attractions occur where the potential energy is minimal or decreasing, and repulsions occur when the potential energy increases. This is an important concept to remember. The potential energy function depicted in Fig. 6-13 is called the *Lennard-Jones potential*, named for John Lennard-Jones who first proposed it in 1924 as a mathematical model to describe two neutral atoms approaching each other.

Hydrophobic Interactions

The highly polar character of water makes for interesting interactions between water and nonpolar substances. The statement "oil and water don't mix" comes to mind. The interaction between water and a nonpolar substance like oil is called the *hydrophobic effect* or *hydrophobic interaction*. The interaction does not involve a single, physical force. Nonetheless *hydrophobic interactions* are sometimes also called *hydrophobic forces*, because the hydrophobic effect forces molecules to arrange themselves in a way that minimizes the contact between water and the nonpolar substance. Thus, water and oil separate, forming two separate phases; this minimizes the amount of contact between oil molecules and water molecules.

The driving force behind hydrophobic interactions is the fact that water "likes" to form hydrogen bonds. Hydrogen bonding between water and itself or between water and other polar molecules lowers the potential energy. It is thus energetically favorable. When a nonpolar or hydrophobic molecule is placed into water, it takes up space between the water molecules restricting their ability to hydrogen bond with each other. There is a price to pay for this in terms of free energy.

First, there is the enthalpy needed to break some of the hydrogen bonds between the water molecules. Second, although hydrogen bonding in water represents a highly ordered, low-entropy state for the molecules (since they must line up with particular orientations to bond with each other), the presence of a nonpolar (hydrophobic) molecule actually decreases the entropy of the water.

Remember order in this context is inversely related to the number of ways of doing something. More order means fewer ways of doing something (less entropy). Less order means more ways of doing the same thing (more entropy). Water molecules by themselves are free to rotate and form hydrogen bonds in many orientations. But, at the surface of a nonpolar or hydrophobic molecule, where the water comes in contact with the hydrophobic molecule, steric interactions restrict the water from rotating freely.

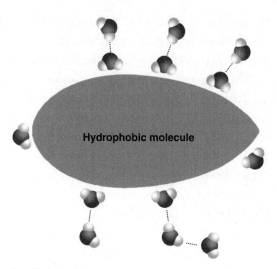

FIGURE 6-14 • Water molecules at the surface of a hydrophobic molecule have their movement limited by steric interactions; this in turn limits the possibilities for hydrogen bonding with other water molecules.

Figure 6-14 illustrates water molecules at the surface of a hydrophobic molecule. The water molecules orient themselves with the hydrogen atoms toward the hydrophobic surface. This orientation allows the water to come as close as possible to the hydrophobic molecule, while at the same time exposing only the most neutral side of the water molecule toward the hydrophobic surface. In this orientation, the surface water is restricted from rotating freely due to steric interactions. This restricted rotation then limits the possible angles for hydrogen bonds to form between the surface water molecules and other water molecules in solution, thus decreasing the entropy of water that is closest to the surface water. This close-to-the-surface water, however, is free to rotate and will do so if it can favorably break the hydrogen bonds with the surface water and make them with other water molecules in solution.

(As a side note, we can explain why this is the most neutral side of the water molecule. First, the partially positive charge on this side of the molecule is split between the two hydrogen atoms, and separated by the large angle between them. This makes the charge effectively weaker in this orientation. Second, the large angle between the hydrogen atoms exposes some of the electronegative oxygen on that side of the molecule as well. This further neutralizes the amount of charge that is exposed on that side of the water molecule.)

Recall Eq. (4-13), shown here as Eq. (6-14).

$$\Delta G = \Delta H - T\, \Delta S \qquad\qquad (6\text{-}14)$$

Both an increase in enthalpy (needed to break some of the water-water hydrogen bonds) and a decrease in entropy (the restricted movement of water molecules at the surface of the hydrophobic molecule) contribute to an increase in the Gibbs energy. Thus mixing of water and hydrophobic substances is energetically unfavorable. This energetic cost can be minimized by minimizing contact between the water and the hydrophobic molecules. In the case of oil and water mixed together, the oil droplets begin to coalesce into larger drops, eventually separating into two phases. The result is a decrease in the Gibbs energy by minimizing contact between the oil and water. So the process of oil and water separating is spontaneous, that is, energetically favored.

The hydrophobic molecules that concern us in biological systems are mostly *hydrocarbons*, molecules made up of only hydrogen and carbon. Hydrocarbons are found abundantly in living systems. Fats and oils are mostly hydrocarbons.

Many biomolecules are not purely hydrophobic, but have a portion of the molecule that is hydrophobic and a portion of the molecule that is *hydrophilic*. *Hydrophilic* means that the molecule interacts favorably with water (lowers the Gibbs energy). Typically a hydrophilic molecule (or the hydrophilic part of a molecule) is highly polar, and easily forms hydrogen bonds with water.

Molecules that are both hydrophobic and hydrophilic are called *amphipathic* or *amphiphilic*. These mean the same thing. (You may occasionally see the incorrect word *amphiphatic* which is a common mistake made by incorrectly combining *amphipathic* and *amphiphilic*.) Phospholipids, which make up the membranes of cells, are amphipathic. Phospholipids have a highly polar (negatively charged) phosphate "head" that is covalently bonded to a long, hydrophobic hydrocarbon "tail." See Fig. 6-15. Many protein molecules also have parts that are hydrophobic and other parts that are hydrophilic.

Hydrophilic Interactions

Hydrophilic structures are those that easily hydrogen bond with water. This creates an attractive force between the hydrophilic structure and water. As mentioned, protein molecules often contain hydrophilic structures. The backbone of DNA and the head groups of phospholipids, because they contain ionized charges, also are also highly hydrophilic. (In general any ion or ionized portion of a molecule is hydrophilic since ions easily form hydrogen bonds with water.)

There are situations, however, in biological systems where two hydrophilic surfaces have to come together, for example, during membrane fusion, in which the membranes of two cells touch and then fuse together, connecting as a single

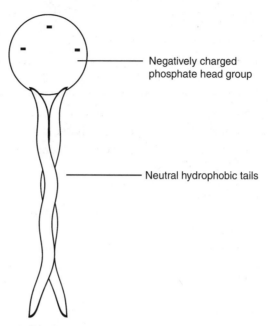

Negatively charged
phosphate head group

Neutral hydrophobic tails

FIGURE 6-15 • Phospholipids are amphipathic; they have a hydrophilic (highly polar) head phosphate group and a hydrophobic (neutral) tail.

membrane. The membrane surface is hydrophilic. Fusing of two membranes must therefore begin with bringing two hydrophilic surfaces together. When a protein folds into a globular shape, sometimes different hydrophilic parts of the protein must also come together. Or when a hydrophilic protein binds to the hydrophilic DNA backbone, again, two hydrophilic surfaces must come together. The problem in all of these situations is that water molecules are tightly bound to the hydrophilic surfaces. This blocks the surfaces from coming closer than about 25 angstroms (Å) unless the water is somehow removed and the hydrogen bonds broken. This can be done, but there is an energetic cost. The presence of the water effectively acts as a close-range repulsive force keeping the two hydrophilic surfaces apart. We call this the *hydrophilic* force even though it is not a single force but the combined effect of multiple hydrogen bonds, steric interactions, and entropy considerations.

Overcoming the hydrophilic force to bring two polar surfaces together can be done through dispersion forces (as is the case with membranes), through displacement of the water with ions, or through stronger ionic and hydrogen bonding between the two hydrophilic molecules (e.g., when a protein binds to DNA). The release of water from a hydrophilic surface increases the entropy of the water which also contributes to overcoming the hydrophilic force.

QUIZ

Refer to the text in this chapter if necessary. Answers are in the back of the book.

1. Which of the following statements about covalent bonds are true?

   S1. Covalent bonds result from sharing of electrons between atoms.

   S2. Covalent bonds are polar if electrons are shared unevenly.

   S3. Single covalent bonds allow free rotation of the atoms or groups of atoms on each side of the bond.

   S4. Double covalent bonds allow free rotation of the atoms or groups of atoms on each side of the bond.

   S5. Covalent bonds may be single, double, or triple, depending on the number of electrons that are shared.

   S6. Covalent bonds are stronger than hydrogen bonds.

   A. S1, S2, S3, S4 and S5
   B. S1, S3, S5 and S6
   C. S1, S2, S3, S5 and S6
   D. S1, S2, S4, S5 and S6

2. The internal energy that results from a charge-charge interaction is proportional to what?

   A. The distance between the charges.
   B. The square of the distance between the charges.
   C. The reciprocal of the distance between the charges.
   D. The square of the reciprocal of the distance between the charges.

3. The internal energy that results from a charge-dipole interaction

   A. Depends on the angle between the dipole axis and a straight line between the dipole and the charge.
   B. Decreases faster than a charge-charge interaction, as the distance between the charge and the dipole increases.
   C. Increases faster than a charge-charge interaction, as the distance between the charge and the dipole increases.
   D. A and C.
   E. A and B.

4. If a dipole is allowed to rotate freely in solution, how does this affect the internal energy of an interaction between the dipole and a charge?

   A. The internal energy becomes stronger, due to thermal averaging.
   B. The internal energy drops off must faster with distance, due to thermal averaging.
   C. Increased temperature decreases the effect of thermal averaging.
   D. A and B.
   E. B and C.

5. **What does the term *cation-pi interactions* refer to?**
   A. Cations interact with aromatic ring electrons, causing them to fluctuate in a pattern that resembles slices of pie within the aromatic ring.
   B. The internal energy due to interactions between cations and aromatic ring electrons can only occur in multiples of pi, or 3.14 J.
   C. The layer of positive charge along the edges of an aromatic ring behaves like a ring-shaped cation.
   D. Delocalized aromatic ring electrons form a layer of negative charge that can interact with nearby cations.

6. **Forces that arise from synchronized fluctuating charge distributions are called what?**
   A. Stacking interactions
   B. Synchronization forces
   C. Charge-flux interactions
   D. Dispersion forces

7. **What are stacking interactions?**
   A. A specific type of charge-flux interaction unique to aromatic ring structures.
   B. Repulsive forces that prevent molecules from stacking up too high before they fall over.
   C. A type of dispersion force that results in an attraction between stacked aromatic ring structures.
   D. Synchronized charged distributions in platelets.

8. **The hydrophobic effect causes water and oil to separate. Why?**
   A. Direct contact between water and hydrophobic substances, such as oil, greatly decreases the entropy of the water. When the oil and water separate, this minimizes their contact, thus minimizing the Gibbs energy cost of restricting the movement of the water molecules.
   B. Direct contact between water and hydrophobic substances, such as oil, is encouraged by the hydrophobic effect. When the oil and water separate, it is easier to see where they are directly in contact with each other.
   C. Oil has a physical property called *hydrophobia*.
   D. The Gibbs energy of mixing oil and water decreases the entropy of the water more than it does of the oil. Separating the oil and water allows nature to reduce the entropy of the water without affecting the oil.

9. **Which of the following statements are true?**
   S1. Hydrophilic molecules easily form hydrogen bonds with water.
   S2. Hydrophobic bonds occur between hydrophobic molecules.

S3. The hydrophilic force accounts for the difficulty in bringing two hydrophilic molecules so close together that water cannot fit in between them.

S4. The hydrophilic force accounts for the attraction between hydrophilic molecules and water.

A. S1, S2, S3, and S4
B. S1, S2, and S4
C. S2 and S3
D. S1 and S4
E. S1 and S3

10. Steric interactions describe what?

A. Repulsive forces between atoms at very close range.
B. Repulsive forces between steroids and similar aromatic compounds.
C. Attractive forces between proteins.
D. Sharing of electrons.

chapter **7**

Biomolecules 101

In this chapter we review the most commonly studied biomolecules: what they are, what elements they are made from, the basics of their structure, and the roles that they play. In the next chapter, we will review the environment in which biomolecules are most commonly found: the cell. These two chapters together will prepare us for a more detailed study of the physics of these amazing molecules.

CHAPTER OBJECTIVES

In this chapter, you will

- Review the most commonly studied biomolecules.
- Learn the basics about the composition and structure of carbohydrates, lipids, proteins, and nucleic acids.
- Learn the various roles that biomolecules play.

Classes of Biomolecules

It makes sense that the most studied biomolecules are the ones that play the most important roles in life. Following is a table of common classes of biomolecules and some of the roles they play.

CHNOPS

CHNOPS is an acronym for the elements most commonly found in living cells (pronounced ch*i*-n*o*ps, with a short *i* as in "chip" and a short *o* as in "top"). CHNOPS stands for Carbon, Hydrogen, Nitrogen, Oxygen, Phosphorus, and

TABLE 7-1 Common classes of biomolecules and their roles.

Class of Molecule	Some Major Roles
Carbohydrates	Energy storage
	Structural material
	Recognition (receptors)
	Components of nucleic acids and some coenzymes
Lipids	Membrane structure
	Energy storage
	Insulation and protection
Proteins	Catalysis (speeding up) of biochemical reactions Proteins that catalyze biochemical reactions are called *enzymes*
	Control or regulation of biophysical processes
	Protection–immunization
	Transport of other molecules across membranes and around the body
	Structural protection for nucleic acids
	Muscle contraction and intracellular movement
Nucleic acids	Storage of genetic information
	Replication of genetic information (i.e., passing to the next generation)
	Expression of genetic information
	Protein manufacture

Sulfur. Living things do contain small amounts of other elements, for example fluorine (Fl) in tooth enamel, iron (Fe) in blood, chlorine (Cl) to help with digesting food, calcium (Ca) for bones and muscle and nerve function, and sodium (Na) also involved in muscle and nerve function. Together CHNOPS accounts for about 98% of the atoms in most living things. For organisms with bones, calcium adds another 1.5% meaning that for those organisms, CHNOPS plus calcium accounts for about 99.5% of all the atoms in the organism.

Biopolymers

As was briefly discussed in Chap. 2, many biomolecules are *polymers*, macro-molecules that are formed by covalently chaining together a series of smaller molecules. The individual smaller molecules that are chained together are called *residues*. When the residues are all identical, the polymer is called a *homopolymer*. When the residues are different, it is called a *heteropolymer*. Typically bio-polymers are heteropolymers. However, although the residues are different, the individual residues making up a given polymer always have something in common, something that makes them all similar to one another. Nearly all of the molecules discussed in this chapter are biopolymers or form biopolymers. So in this chapter we will get to see some specific examples of biopolymers.

Hydrocarbons and Hydrocarbon Chains

A *hydrocarbon* is a molecule made of only hydrogen and carbon. Many biomol-ecules either are hydrocarbons or large portions of the molecule are hydrocar-bons. A *hydrocarbon chain* is a hydrocarbon in which the carbon atoms are covalently connected to one another in a line, like links in a chain. Each carbon atom has anywhere from zero to three hydrogen atoms attached to it.

The covalent bonds between any two carbon atoms may be single, double, or even triple bonds, indicating one, two, or three electron pairs shared between the atoms. Figure 7-1 shows some example hydrocarbons. Two of the examples in Fig. 7-1 are hydrocarbon chains. One is an aromatic hydrocarbon (see the previ-ous chapter for the definition of aromatic molecules). Many biomolecules consist of hydrocarbon chains with various other atoms attached to the hydrocarbon.

Carbon Covalent Bonds

While we are on the topic of hydrocarbons, this is a convenient time to talk about the nature of chemical bonding with carbon atoms and some convenient and important rules to remember.

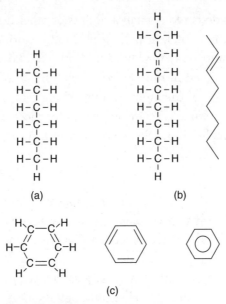

FIGURE 7-1 • Example hydrocarbons.
(a) Hexane. (b) 2-Octene. Shown using a structural formula (left) and skeletal formula (right). Notice there is one double bond between the first and second carbon.
(c) Benzene, also shown using structural and skeletal formulas. The third diagram on the right shows the skeletal formula with a circle in the middle. This is a special way of indicating aromatic rings in skeletal formulas so as to indicate that the bonding electrons are delocalized (see Chap. 6).

Rule 1. Carbon atoms have four valence electrons to share. Therefore, in a molecule, carbon atoms will form four covalent bonds with other atoms.

It is this rule that allows us to write skeletal formulas for hydrocarbons and other biomolecules as in Fig. 7-1. A skeletal formula is a short cut way of showing a molecular structure. A skeletal formula shows each carbon covalent bond as a line segment, except that the covalent bonds directly between carbon atoms and hydrogen atoms are not shown (as a shortcut). A skeletal formula also does not show the carbon atoms themselves, nor does it show the hydrogen atoms that are directly connected to the carbon atoms.

By this definition of a skeletal formula, each carbon atom is positioned where any two line segments meet (i.e., where there is sharp bend or angle between two line segments in the skeletal formula) or wherever a line segment comes to an end.

Also, by this definition, hydrogen atoms connected directly to carbon atoms are not shown. So how do we know how many hydrogen atoms are connected to each carbon atom? This is not a problem because we can use Rule 1. For example, where two line segments meet, we know we have a carbon atom with two covalent bonds to atoms other than hydrogen. Since Rule 1 tells us that carbon forms four covalent bonds, we know that there must be two hydrogen atoms covalently bonded to that carbon. As another example, at the end of the chain, where a single line segment comes to a point, we again use Rule 1 that each carbon has four covalent bonds. Since only one bond is shown (the last line segment), we know that the end carbon atom must have three hydrogen atoms attached. You can see this in Fig. 7-1 by comparing the skeletal formulas with the corresponding structural formula.

Sometimes carbon atoms form double bonds. A *double bond* involves sharing two electron pairs (i.e. four electrons) in the same covalent bond between two atoms. In the structural and skeletal formula, double bonds are shown as two parallel line segments. Sometimes, but less often, carbon atoms even form triple bonds. As you might expect, a *triple bond* involves sharing three electron pairs (i.e., six electrons) in the same covalent bond between two atoms, and a triple bond is shown as three parallel line segments in structural and skeletal formulas.

> *Rule 2.* When counting covalent bonds for a carbon atom, double bonds count as two covalent bonds and triple bonds count as three covalent bonds.

Sometimes we need to use Rule 2 in conjunction with Rule 1 to determine how many hydrogen atoms are directly connected to a carbon atom. You can see an example of this in Fig. 7-1. Notice that where there are some carbon atoms that are shown to have a double bond on one side and a single bond on the other side. That makes three covalent bonds shown for each of those carbon atoms. Since each carbon atom must have four covalent bonds, we know that those particular carbon atoms each have only one hydrogen atom directly attached.

Functional Groups

A *functional group* is a specific group of atoms within a molecule that confers certain characteristics to that molecule. There are several functional groups that are quite common in biomolecules. To help us speak about of the composition and structure of various biomolecules, it will be helpful to first review some common functional groups.

Hydroxyl Group

A *hydroxyl* group is an oxygen atom with one hydrogen atom covalently attached. The hydrogen atom is attached only to the oxygen atom, whereas the oxygen atom attaches the hydroxyl group to the rest of the molecule. Figure 7-2 shows a hydroxyl group. The R in the figure indicates the rest of the molecule. The rest of the molecule can be one, several, hundreds, or even thousands of atoms. The oxygen atom of the hydroxyl group, however, is always bonded to only one atom of the rest of the molecule. In biomolecules the hydroxyl oxygen is most often bonded to a carbon atom. But it can be attached to other atoms, for example, to phosphorous, in the case of a *phosphate* functional group. Hydroxyl groups are polar, with the hydrogen atom having a slightly positive charge allowing it to participate in *hydrogen bonding* with other atoms (see Chap. 6).

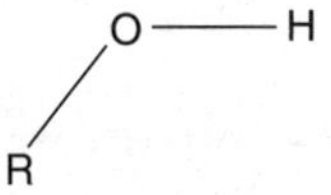

O———H
/
R

FIGURE 7-2 • A hydroxyl group is an oxygen atom covalently bonded to a hydrogen atom. In a hydroxyl group, the hydrogen is bonded only to the oxygen, whereas the oxygen atom connects the hydroxyl group to the rest of the molecule.

Carbonyl Group

A *carbonyl* group is carbon atom with a double covalent bond to an oxygen atom. In addition to the double bond with the oxygen, the carbon atom also has two covalent single bonds to other atoms. At least one of these single bonds is always connected to the rest of the molecule. There are various types of carbonyl groups depending on what atom or functional group is connected to each of the two single bonds. Figure 7-3 shows three common types of carbonyl

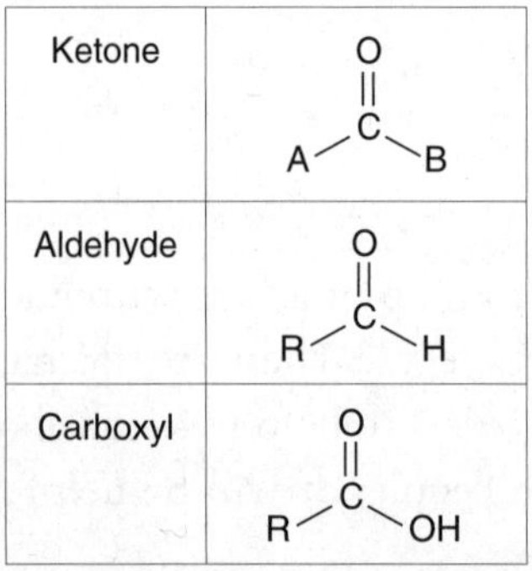

Ketone	$\overset{\displaystyle O}{\underset{A^{\diagup}\overset{\displaystyle \|}{C}\diagdown B}{}}$
Aldehyde	$\overset{\displaystyle O}{\underset{R^{\diagup}\overset{\displaystyle \|}{C}\diagdown H}{}}$
Carboxyl	$\overset{\displaystyle O}{\underset{R^{\diagup}\overset{\displaystyle \|}{C}\diagdown OH}{}}$

FIGURE 7-3 • Three types of carbonyl groups.

groups. When both single bonds are connected to the rest of the molecule, the group is called a *ketone*. When one of the two single bonds is connected only to a hydrogen atom, then the group is called an *aldehyde*. When it is connected to a hydroxyl group, then the carbonyl is called a *carboxyl* group.

The double bond between the carbon and oxygen is a polar bond, so the oxygen has a slightly negative charge and the carbon has a slightly positive charge. Carbonyl oxygen atoms often participate in hydrogen bonding.

Carboxyl Group

A *carboxyl* group is a type of carbonyl group. We discuss it here in a section of its own because it is so common in biomolecules. In a carboxyl group, the carbon atom has a double covalent bond to an oxygen atom and a single covalent bond to a hydroxyl group. The remaining single covalent bond attaches the carboxyl group to the rest of the molecule (see Fig. 7-4a).

The chemical formula for a carboxyl group is usually written as COOH. Molecules that contain a carboxyl functional group are called *carboxylic acids*. The hydrogen atom of the carboxyl group has a slightly positive charge, and the doubly bonded oxygen has a slightly negative charge, making them both able to participate in hydrogen bonding (see Fig. 7-4b). Carboxyl groups are also easily ionized to form COO^- and H^+ ions.

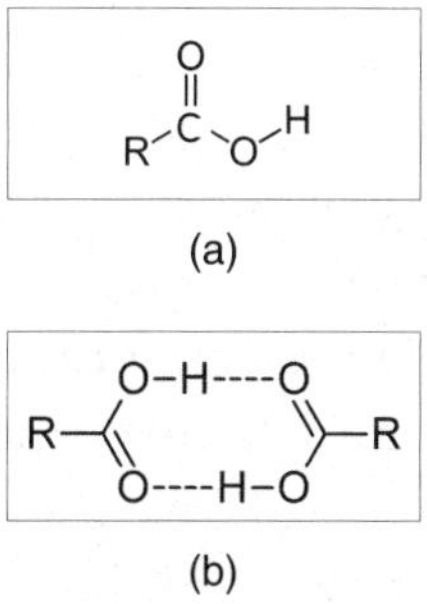

(a)

(b)

FIGURE 7-4 · A carboxyl group is carbon atom doubly bonded to an oxygen atom, and singly bonded to a hydroxyl group. (a) Molecules containing carboxyl groups are called *carboxylic acids*. (b) In carboxyl groups, both the doubly bonded oxygen and the hydroxyl hydrogen are able to participate in hydrogen bonding. The figure shows two carboxylic acids hydrogen bonding to each other.

Amine Group

An *amine* group is a nitrogen atom bonded in such a way as to have a pair of nonbonded valence electrons. For our purposes, you can think of an *amine* as a

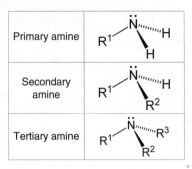

FIGURE 7-5 · An amine group is a nitrogen atom with a pair of nonbonded valence electrons. The amines that concern us most are primary and secondary amines which have two and one hydrogen atom, respectively.

nitrogen atom that has three single covalent bonds to three other atoms. Typically two of the other atoms are hydrogen, while the third atom is a carbon atom linking the amine group to the rest of the molecule. This is called a *primary amine*, because it has one connection to the rest of the molecule. See Fig. 7-5. A *secondary amine* has only one hydrogen atom attached and two covalent bonds to the rest of the molecule. Tertiary amines exist but are less common. In a *tertiary amine* the nitrogen has three connections to the rest of the molecule, with no hydrogen atoms directly connected to the nitrogen.

Amine groups are polar with a slightly negative charge on the nitrogen atom and (for primary and secondary amines) a slightly positive charge on the hydrogen atoms. Amines can participate in hydrogen bonding, on either the positive or negative side of the bond. One amine group can thus participate in up to three hydrogen bonds at once. Amines can form positive ions by covalently binding to an extra hydrogen atom without its electron (i.e., by covalently bonding to a proton). Thus, when ionized, we have NH_3^+ for a primary amine ion, NH_2^+ for a secondary amine ion, and NH^+ for a tertiary amine ion.

Phosphate

A *phosphate* group is a phosphorous atom with four oxygen atoms attached. One oxygen atom always shares a double bond with the phosphorous. The remaining three oxygen atoms either are covalently bonded to the rest of the molecule or are part of hydroxyl groups. Figure 7-6 shows a phosphate group with only one oxygen atom connecting it to the rest of the molecule. Thus two of the four oxygen atoms are part of hydroxyl groups. Phosphate groups are highly polar and so can form hydrogen bonds, as we

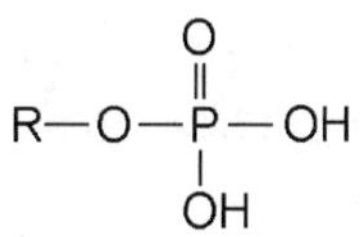

$$R-O-\underset{\underset{OH}{|}}{\overset{\overset{O}{\|}}{P}}-OH$$

FIGURE 7-6 · A phosphate group is a phosphorous atom with four oxygen atoms attached. One oxygen atom always shares a double bond with the phosphorous. The remaining three oxygen atoms either are covalently bonded to the rest of the molecule or are part of hydroxyl groups.

saw with the other functional groups discussed. Phosphate groups also easily form negative ions by losing the hydrogen atom from one or both of its hydroxyl groups.

Acids, Bases, and pH

Aqueous solutions always contain some hydrogen ions (H^+) and hydroxyl ions (OH^-). Since a hydrogen atom is just one proton and one electron, a hydrogen cation (H^+) is really just a proton in solution. Molecules or functional groups that increase the concentration of protons in solution are called *acids*. Those that decrease the concentration of protons are called *bases*.

For example, amines easily bind protons (because of their nonbonded valence electrons). Binding protons reduces the proton concentration in solution, and so amines are considered *basic*. Phosphates easily release protons and so phosphates are considered *acidic*. Sometimes bases decrease the concentration of protons by increasing the concentration of hydroxyl ions (OH^-). The hydroxyl ions bind to protons and thereby decrease the concentration of protons.

We measure the concentration of protons in units of pH (pronounced "pee-aych"). The pH is approximately equal to the negative base-ten logarithm of the proton concentration. A pH of 7 means H^+ is at a concentration of 10^{-7} moles per liter. Pure water has a pH of 7 and is considered neutral. A pH less than 7 means a higher concentration of H^+ (e.g., pH = 3 means 10^{-3} moles per liter which is greater than 10^{-7} moles per liter). Acidic substances (those that increase H^+) therefore *lower* the pH, whereas bases *increase* pH. Since water, with a pH of 7, is considered neutral, measurement of pH is considered to range from pH = 0 for the most acidic substances to pH = 14 for the most basic substances.

Many biochemical and biophysical processes are sensitive to the concentration of protons and hydroxyl ions. These ions also affect the forces that determine the conformation of biomolecules, especially where other ions or hydrogen bonds are involved.

We've just finished learning the main functional groups that are significant in biomolecules, as well as some other important background information relevant to our understanding of molecular and subcellular biophysics. Next we use this information to discuss the four major classes of biomolecules: carbohydrates, lipids, proteins, and nucleic acids.

Still Struggling

The bulk of this chapter is all about nomenclature, building up a vocabulary so that we can talk about the physics of biomolecules and how it relates to their function. You may need to review this chapter more than once if the names don't stick right away. To make it easier, focus on the section headings. These are the main terms that you need to know. Look over the figures. You don't need to memorize the chemical structures, but you should know what elements are in them and what that means from a physical point of view. For example, you should know that hydrocarbons contain only hydrogen and carbon. And molecules, or portions of molecules, that are purely hydrocarbon are hydrophobic (avoid water). On the other hand, functional groups that contain oxygen and nitrogen tend to be hydrophilic, and may participate in hydrogen bonding. Acids and bases also tend to be hydrophilic because they easily ionize. Structures that contain aromatic rings are relatively stiff and inflexible. Above all, use this chapter as a reference. When you see these terms in later chapters, there will usually be enough information to jog your memory, but if not simply return here.

Carbohydrates

Carbohydrates are molecules made of carbon, hydrogen, and oxygen. They are the most abundant biomolecules on Earth. Carbohydrates are used for energy storage (food). They also serve as structural material, for example, cellulose. Carbohydrates are components of nucleic acids (the genetic material) and of some coenzymes. Carbohydrates also play a role in recognition when one cell or molecule needs to recognize another.

In biochemistry, carbohydrates are also called *saccharides* and in common language *sugars*. The simplest of all carbohydrates are *monosaccharides*, simple, single-unit carbohydrates. The chemical formula of a monosaccharide is

$$C_nH_{2n}O_n$$

This formula tells us that monosaccharides contain equal amounts of carbon atoms and oxygen atoms, and twice as many hydrogen atoms. Monosaccharides contain from 3 to 8 carbon atoms. Figure 7-7 shows the structure of a simple monosaccharide, glucose.

When two monosaccharides are covalently bonded together into a single molecule, the molecule is called a *disaccharide*. Sucrose, or table sugar, is a common disaccharide. When a carbohydrate polymer contains between 2 and 20 monosaccharides, we call it an *oligosaccharide*. We usually reserve the term *polysaccharide* to refer to carbohydrate molecules containing greater than

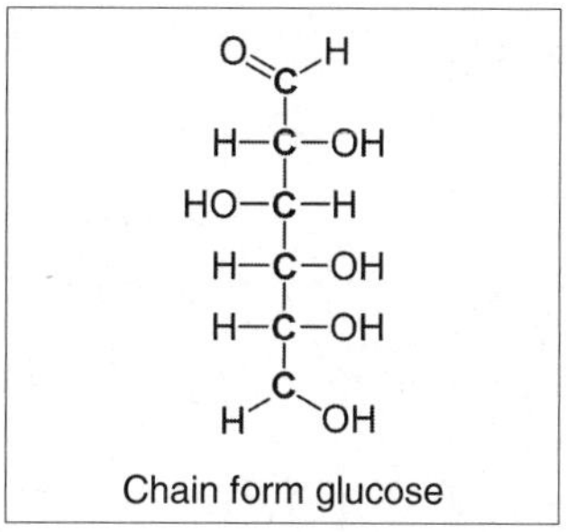

Chain form glucose

(a)

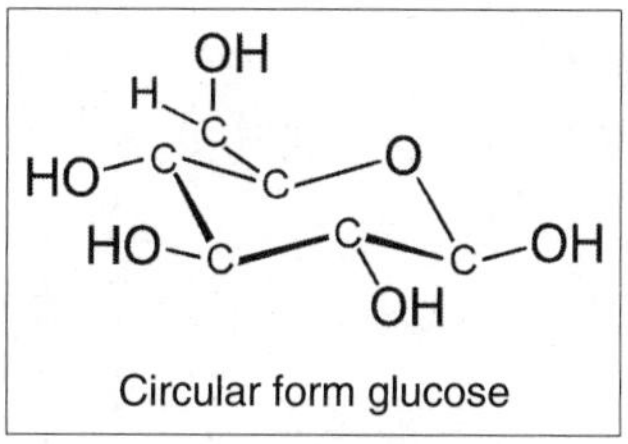

Circular form glucose

(b)

FIGURE 7-7 · The monosaccharide glucose has the chemical formula $C_6H_{12}O_6$. Two structures of glucose are shown. (a) Chain form glucose. (b) Circular form glucose. The circular form is most common in biological systems.

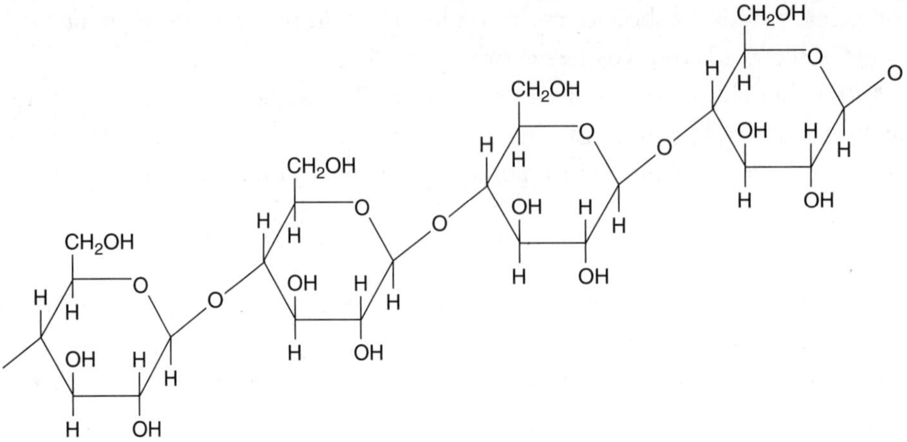

FIGURE 7-8 • The polysaccharide cellulose is composed of glucose residues. (*Courtesy of Biochemistry Demystified.*)

20 monosaccharides. Although all saccharides are commonly called sugars, polysaccharides are also called *starches*. Figure 7-8 shows a portion of the polysaccharide cellulose, which is composed of glucose residues linked together.

Derivatives of Carbohydrates

Some molecules, strictly speaking, are not classified as carbohydrates, but are derived from and are similar to carbohydrates. For example, if a given carbohydrate molecule has one less oxygen atom than normally found in carbohydrates, so that its chemical formula is $C_nH_{2n}O_{n-1}$, we call it a *deoxycarbohydrate*. The most studied deoxycarbohydrate is *deoxyribose* ($C_5H_{10}O_4$), which is a major component of the residues making up DNA (deoxyribonucleic acid).

Other derivatives of carbohydrates include *sugar alcohols* such as sorbitol, which are used as low-calorie sweeteners. (They also reduce tooth decay because, unlike sugars, they cannot be digested by the bacteria in the mouth.) *Sugar acids* are also derived from carbohydrates. Ascorbic acid, or vitamin C, is one such example.

Lipids

Lipids as a class of biomolecules do not dissolve in water, but they do dissolve in nonpolar solvents. Lipid molecules are all either hydrophobic or amphipathic. Some lipids are polymers, or may form polymers. Other lipids are never found as polymers.

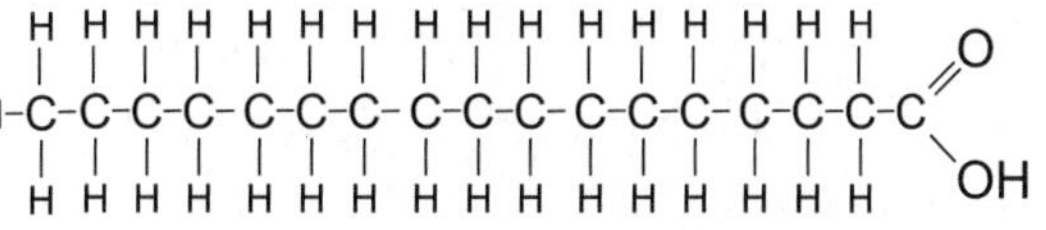

(a)

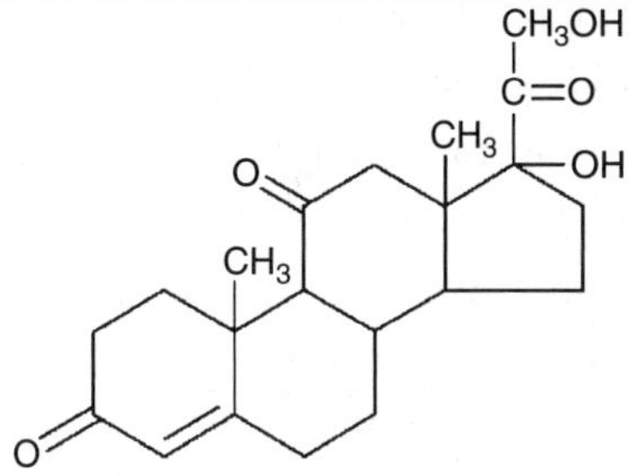

(b)

FIGURE 7-9 · Example structures of two classes of lipid: polymeric and nonpolymeric. (a) The fatty acid palmitic acid (a major component of palm tree oil). Fatty acids are long chain hydrocarbons with a carboxyl group at one end. (b) Cortisone, a steroid. Steroids are a class of nonpolymeric lipids characterized by four covalently attached ring structures.

Probably the most important job that lipids do is to serve as the main structure of cell membranes. Membranes protect and compartmentalize the insides of cells. They also help decide what gets into and out of the cell. Another role for lipids is energy storage. Excess energy from the food we eat can be stored in the chemical bonds that make up a lipid polymer. Later, this energy can be released and used by the organism. Lipids, in the form of stored fat, also provide a layer of thermal insulation and protection for many organisms.

Figure 7-9 shows an example of a polymeric lipid and nonpolymeric lipid. As you can see, the two categories of lipid are significantly different in structure. Polymeric lipids typically contain long chain hydrocarbon structures. Nonpolymeric lipids often contain one or more ring structures. What they have in common, what makes them both lipids, is that they do not dissolve in water but they do dissolve in nonpolar solvents.

Fatty Acids

Fatty acids are long chain hydrocarbons with a carboxyl (carboxylic acid) group at one end. Fatty acids can be characterized in terms of the number of carbon atoms in the chain, which is typically from 12 to 20 carbon atoms. Fatty acids are also characterized in terms of whether any double bonds exist in the hydrocarbon chain and, if so, where the double bonds are in the chain.

Fatty acids that have *no* double bonds are called *saturated* fatty acids. This is because with no double bonds, the carbon atoms are able to covalently bond to the maximum number of hydrogen atoms. Thus they are saturated with hydrogen atoms. As explained, carbon atoms form four covalent bonds. A carbon atom in the middle of the chain can bond to at most two hydrogen atoms, since it is also bonded to two other carbon atoms (one on each side in the chain). The carbon atom at the *end* of the chain is covalently bonded to three hydrogen atoms, since it is connected to only one other carbon atom.

When a carbon atom is double-bonded with another atom, then the double bond counts as two of the four covalent bonds that the carbon atom can form. If the double-bonded carbon atom is in the middle of chain, then it can bond with only one hydrogen atom, since it otherwise has a double bond on one side and a single bond on the other. If the double-bonded carbon is at the end of the chain, then it can only have two hydrogen atoms directly attached to it, instead of its usual three. Fatty acids that contain double bonds are called *unsaturated*, because it is possible (through chemical processes) to add hydrogen atoms to the molecule by converting the double bonds to single bonds. Fatty acids with more than one double bond in the carbon chain are called *polyunsaturated*.

Glycerides

Glycerol is a small, three-carbon carbohydrate. Glycerol can be derived from a saturated, three-carbon hydrocarbon chain by replacing one hydrogen atom on each of the three carbon atoms with a hydroxyl group. See Fig. 7-10. (Glycerol is also called *glycerin* or *glycerine*; all three terms refer to the same molecule.)

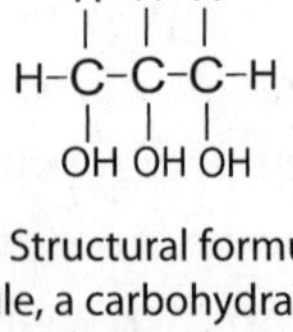

$$
\begin{array}{ccc}
H & H & H \\
| & | & | \\
H-C-C-C-H \\
| & | & | \\
OH & OH & OH
\end{array}
$$

FIGURE 7-10 · Structural formula for a glycerol molecule, a carbohydrate that serves as the foundation for lipids known as glycerides.

FIGURE 7-11 • A triglyceride is a glycerin molecule attached to three fatty acids. The carboxylic acid functional group has lost its hydroxyl hydrogen and replaced it with a covalent bond to the glycerin molecule. In this way the carboxylic acid group has become an *ester*. (*Courtesy of* Biochemistry Demystified.)

Glycerol serves as the foundation for a class of lipids known as *glycerides*. Glycerides are formed by combining one, two, or three fatty acids with the glycerol molecule, resulting in a monoglyceride, diglyceride, or triglyceride.

Triglycerides play an important role in energy storage in many organisms, as well as in dietary intake of lipids. When a fatty acid combines with glycerol, the hydrogen atom on one of the glycerol hydroxyl groups is replaced by a covalent connection to the carboxylic carbon of the fatty acid. At the same time, the fatty acid loses its hydroxyl group. (The lost hydroxyl group combines with the hydrogen atom from glycerol to form a water molecule.) The result is an *ester* functional group attaching the glycerol to the fatty acid hydrocarbon chain. An *ester* is a type of carbonyl group very similar to carboxyl, but where the hydroxyl hydrogen or the carboxyl group has been replaced with a carbon atom. That carbon atom in turn may be attached further to other atoms in the molecule. Figure 7-11 shows an example of a triglyceride. Notice the oxygen atom between the carbonyl carbon and the glycerin portion of the molecule.

Phospholipids

A *phospholipid* is class of lipid consisting of phosphate group attached to a diglyceride. The attachment is via the diglyceride's third glycerol hydroxyl group. This hydroxyl group is replaced by a covalent bond to one of the phosphate oxygen atoms. Typically the other side of the phosphate is attached to some other small organic molecule or functional group. Figure 7-12 shows as an example the phospholipid *phosphatidylcholine*, where the small molecule choline is attached to the phosphate. A single type of phospholipid such as phosphatidylcholine can represent a whole set of phospholipids depending on which specific fatty acid side chains are attached.

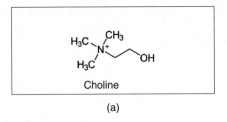

Choline

(a)

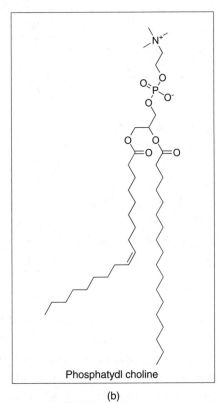

Phosphatydl choline

(b)

FIGURE 7-12 • A choline molecule (top), and the phospholipid phosphatidylcholine (bottom).

Phospholipids are a major component of cell membranes. The phosphate "head" of phospholipids is very hydrophilic, while the hydrocarbon "tails" are hydrophobic. The amphipathic character of phospholipids makes them ideal for membrane construction.

Nonpolymeric Lipids

Nonpolymeric lipids are lipids that are not polymers or do not contain polymers as a major portion of their structure. They are also called *nonsaponifiable* lipids,

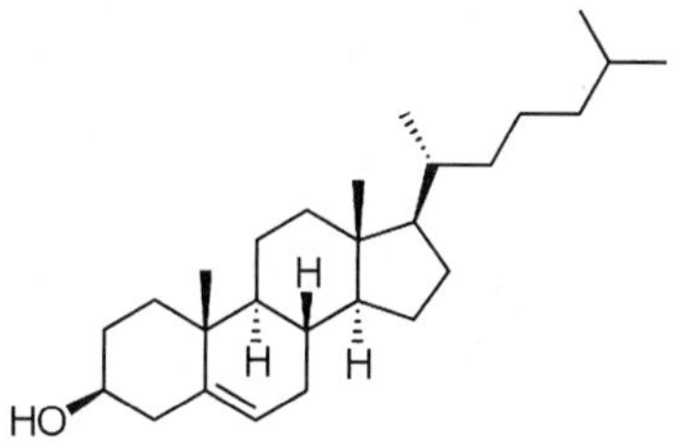

FIGURE 7-13 · Cholesterol.

meaning that they cannot be broken down by the chemical process of hydrolysis. Nonpolymeric lipids typically include one or more ring structures. Examples of nonpolymeric lipids include the steroids (such as cortisone and vitamin D). Other examples include vitamins A, E, and K which are part of a class of molecules known as terpenes. Probably the most famous of nonpolymeric lipids is cholesterol, which is a sterol, a combination of a steroid and an alcohol. See Fig. 7-13. Cholesterol plays an important role in the cell membrane. It is amphipathic. Its hydroxyl group can bind to the hydrophilic portion of phospholipids, and its relatively stiff hydrophobic rings lend rigidity to the hydrophobic portion of cell membranes.

Proteins

A *protein* is a polymer of amino acids, so to understand what a protein is, one first needs to understand what an amino acid is. An *amino acid* is a molecule that has a carbon atom with an amine group on one side and a carboxyl group on the other. This carbon atom is called the *α-carbon* (alpha carbon) to distinguish it as the place where the amine and carboxyl groups are attached. Since carbon atoms typically form four covalent bonds, that leaves two more covalent bonds for the α-carbon. One of those two bonds is formed with hydrogen. The other is formed with a group of atoms called the *amino acid side chain*. The side chain can vary greatly from one amino acid to another. The word *chain* in this context is not meant to imply a linear chain (as it did in the case of a hydrocarbon chain). Rather an *amino acid side chain* is simply a group of a few to about 20 or so atoms making up a specific structure. We call it a *side chain* by convention, because it hangs from the part of the molecule that defines the molecule as an amino acid. See Fig. 7-14.

There are 20 different major amino acids found in living things, each with its own characteristic side chain. Some side chains are hydrophobic. Some are

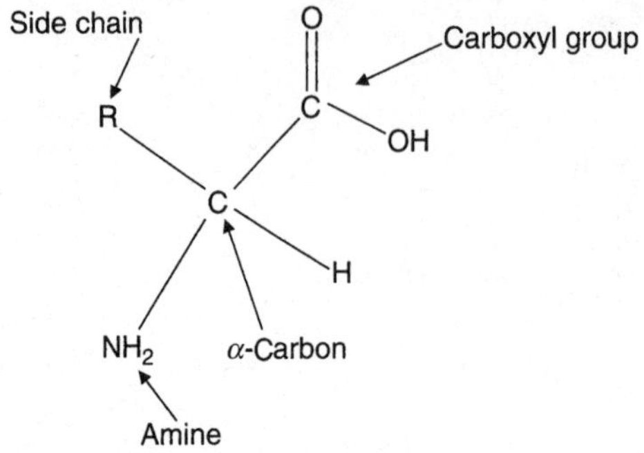

FIGURE 7-14 · Structure of an amino acid.
(*Courtesy of* Biochemistry Demystified.)

hydrophilic. Some are amphipathic. From this group of 20 amino acids, organisms build literally thousands and thousands of different proteins. The extremely large number of different proteins is possible because, like the letters in an alphabet used to form words and sentences, organisms can link together various amounts of 20 different amino acids in various orders to form proteins. When you consider that any given protein can contain anywhere from 15 or so amino acids, to hundreds or even thousands of amino acids, linked together in various arrangements, the number of possible combinations is enormous. The number of different proteins is analogous to the number of different sentences, or perhaps even the number of different paragraphs, one could possibly write given an alphabet.

When amino acids are connected together to form a polymer, the carboxyl group of one amino acid connects to the amine group of the next amino acid. The process is similar to what we saw above when connecting the carboxyl of a fatty acid to glycerol. In the amino acid case, a hydrogen atom from the amine breaks away and combines with the hydroxyl group from the carboxyl, which also breaks away. These two combine to form a new water molecule. At the same time a covalent bond is formed between the carbonyl carbon of one amino acid and the amine nitrogen of another. This bond is called a *peptide bond*. For this reason, proteins are also called *polypeptides*. Figure 7-15 illustrates the formation of a peptide bond between two amino acids.

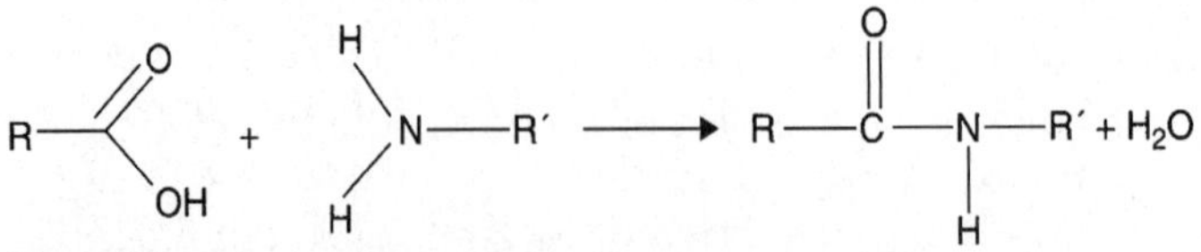

FIGURE 7-15 · Formation of a peptide bond connecting two amino acids. (*Courtesy of* Biochemistry Demystified.)

Protein Structure and Function

Proteins exhibit primary, secondary, tertiary, and quaternary structures (as was discussed in Chap. 2). Once a polypeptide is formed, the primary structure or sequence of amino acids in the polymer bends, twists, and folds, until the protein takes on a particular shape. One of the big questions in biophysics is how can we predict the structure, and possibly function, of a protein from the knowledge of its amino acid sequence. When proteins fold into shape, the most favorable shape is the one that minimizes contact between hydrophobic and hydrophilic amino acids. In aqueous solution, as a general rule, the hydrophobic parts of the protein will tend to be folded into the center of the molecule, shielded by the hydrophilic portions of the protein which have more direct contact with water.

Proteins are the worker bees of life. Proteins take part in nearly every process in living organisms. They regulate biophysical processes, they generate movement, they are essential for nerve signals, they provide structural support, and they transport other molecules from place to place.

Proteins are most famous for catalyzing (increasing the rate of) biochemical reactions. A *catalyst* is a substance that modifies the rate of a chemical reaction, without actually being consumed by the reaction. Proteins that act as catalysts are called *enzymes*. The rate of nearly every biochemical process is controlled by enzymes. One cannot overstate the importance of the role of proteins as enzymes. Even the organism's manufacture of a single protein (or other biomolecule) involves the work of many different proteins acting as enzymes.

In addition to controlling the rate of biochemical reactions, proteins are very much involved in other ways in the control or *regulation* of biological processes. For example, proteins bind to other molecules, and through that binding they either encourage or hinder various biophysical processes. The binding is typically through a set of hydrogen bonds and salt bridges between the protein and the other molecule. There are many mechanisms through which binding affects biophysical processes, including steric hindrance (getting in the way) and stabilizing a particular conformation of the molecule.

Proteins play a major role in the immune system too. Antibodies, also called *immunoglobulins*, are proteins that can recognize and bind to foreign objects (such as viruses and bacteria). In this way they help neutralize the threat from foreign objects and prevent them from harming the organism.

Proteins are involved in transport of other molecules. For example, the protein hemoglobin transports oxygen from the lungs to every cell in the body. Similarly *lipoproteins* carry lipids to where they are needed, often carrying them through places where the lipids would otherwise (without the protein) be insoluble (e.g., through the bloodstream). Proteins also help transport ions and other small molecules through cell membranes.

Proteins play a structural role as well. Examples include holding DNA together in chromosomes (see the section Nucleic Acids that follows), formation of hair, fingernails, and muscles, and formation of the cytoskeleton, a fibrous protein network that helps cells maintain their shape. Proteins also provide movement; for example, muscle contraction is a direct result of the action of proteins, as is movement of the cytoskeleton.

Nucleic Acids

Nucleic acids are biomolecules named for being an *acidic* substance (one that increases the concentration of H^+ ions) originally discovered in the *nucleus* (center compartment) of cells. Nucleic acids play a number of roles in life, including storage and expression of genetic information, transport of amino acids during protein synthesis (manufacture), and in some cases even acting as catalysts.

Nucleic acids are polymers of nucleotides. A *nucleotide* is molecule that has the following three parts to it: (1) a *nitrogenous base* (also called a *nucleotide base*) which is any one of certain aromatic ring compounds containing nitrogen, (2) a sugar (*ribose* or *deoxyribose*), and (3) one to three phosphate groups.

There are five main nitrogenous bases found in nucleic acids. Small amounts of other nucleotide bases do occur, but they are far less common and so will not be discussed here. The five main nucleotide bases fall into two categories: *pyrimidines* and *purines*. The pyrimidines are so named because they are all similar to the molecule *pyrimidine* which is an aromatic ring containing four carbon and two nitrogen atoms. (See Fig. 7-16a.) The pyrimidines include cytosine, thymine, and uracil. In contrast to pyrimidines, a *purine* contains *two* rings; one is a five-atom ring with two nitrogen atoms, the other is a six-atom pyrimidine ring. The purine's two rings are covalently attached to one another in a way such that they share two carbon atoms. That is, two carbon atoms are part of both rings. (See Fig. 7-16b.) The purines include adenine and guanine.

The sugar within the nucleotides is either *ribose* or *deoxyribose*. Both are five-carbon carbohydrates, in ring form (actually strictly speaking the deoxy form is a carbohydrate derivative, as mentioned earlier in this chapter). Figure 7-17 shows the skeletal formula for ribose and deoxyribose. Nucleic acids that contain ribose are called *ribonucleic acids*, abbreviated RNA, and nucleic acids containing deoxyribose are called *deoxyribonucleic acid*, or DNA. There are a number of differences

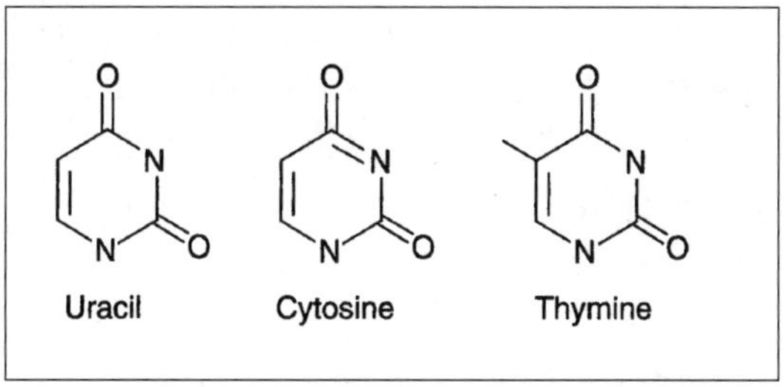

FIGURE 7-16 • The nucleotide bases. There are five main nucleotide bases found in nucleic acids. They are divided into two main categories: (a). Pyrimidines, including cytosine, thymine, and uracil. (b) Purines, including adenine and guanine. (*Courtesy of* Biochemistry Demystified.)

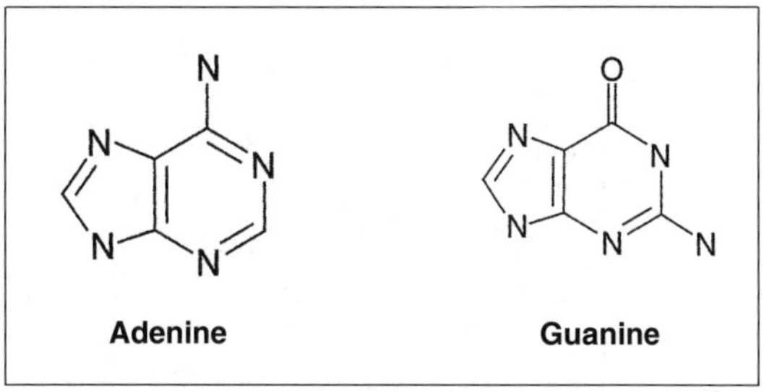

between RNA and DNA, but the two main differences that you should remember are (1) RNA contains the sugar ribose, whereas DNA contains deoxyribose and (2) in RNA we find the nucleotide base uracil (but generally not thymine), whereas in DNA we find thymine (but not uracil).

Within nucleic acids, the nucleotides have only one phosphate group. However, nucleotides play other biological roles, outside of nucleic acids, in which they sometimes have two or three phosphate groups in a row.

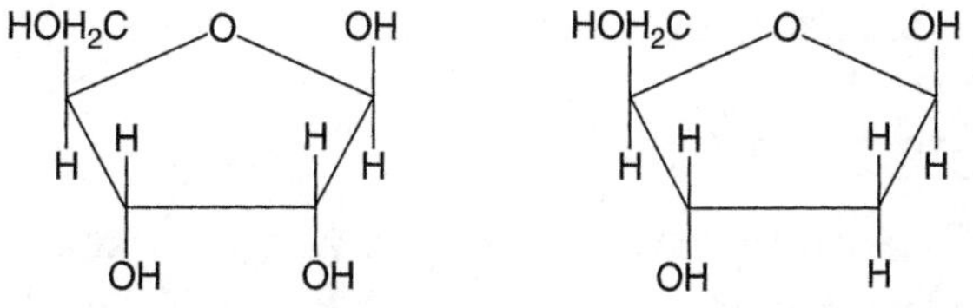

FIGURE 7-17 • (a) Ribose. (b) Deoxyribose.

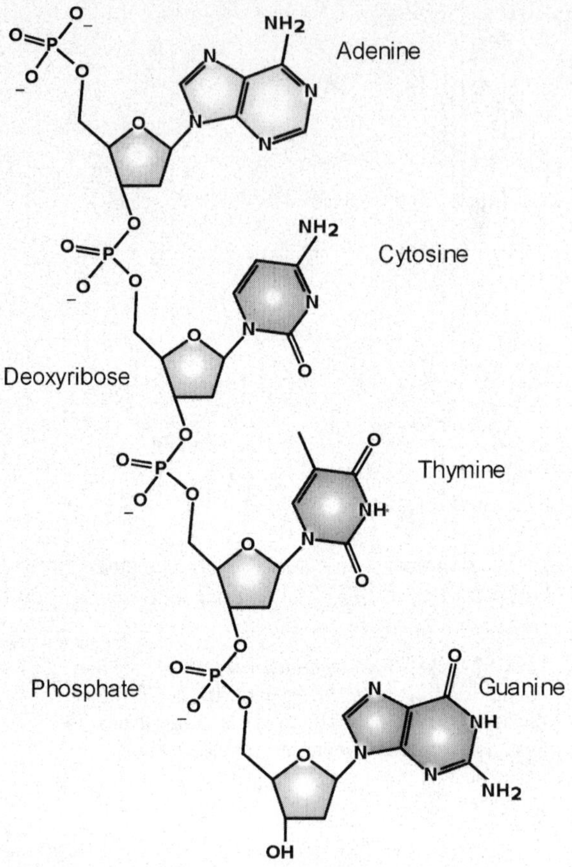

FIGURE 7-18 • An example of a four-residue DNA polymer. (*Adapted from Wikimedia Commons.*)

When nucleotides polymerize to form a nucleic acid, the sugar from one residue covalently attaches to the phosphate from the next residue, and so on. The resulting nucleic acid polymer consists of a backbone of alternating sugar and phosphate residues, with the nitrogenous bases covalently attached to, and hanging off the side of, the sugar residues. See Fig. 7-18.

Nucleic Acid Structure

Nucleic acids (DNA and RNA) exhibit a number of primary, secondary, tertiary, and even quaternary structures that will be discussed in some detail in a later chapter. Here we will provide a brief review of some of the common structural features of nucleic acids.

ACCACGCUUAAGACACCUAGC

FIGURE 7-19 • A portion of the nucleotide
sequence from a molecule of phenylalanine
transfer RNA (a ribonucleic acid). When
specifying the nucleotide sequence (primary
structure) of a nucleic acid, we abbreviate using
the first letter of each nucleotide base: A, T, C,
G, and U, respectively for adenine, thymine,
cytosine, guanine, and uracil.

The primary structure of nucleic acids is the specific *nucleotide sequence* making up the polymer. When discussing primary structure in DNA and RNA, we abbreviate the nucleotides using the first letter from the name of the nitrogenous base, in upper case. That is, we use A, T, C, G, and U, respectively for adenine, thymine, cytosine, guanine, and uracil. Figure 7-19 shows, as an example, the nucleotide sequence for a portion of an RNA molecule known as *phenylalanine transfer RNA*. Phenylalanine is an amino acid. Phenylalanine transfer RNA (abbreviated *tRNA^{Phe}*) is a ribonucleic acid involved in transporting phenylalanine to where it is needed for protein synthesis (manufacture).

Nucleic acid secondary structures tend toward a helical shape, largely due to aromatic stacking interactions (see Chap. 6) among the nitrogenous bases. Biophysicists use the phrase *base stacking*, when referring specifically to the stacking interactions of nitrogenous bases in nucleic acids, but it is the same aromatic stacking discussed in Chap. 6.

Nucleic acids form both *single-stranded* and *double-stranded polymers*. Both are helical structures. In the case of double-stranded nucleic acids, the secondary structure is the famous *double helix*. The double-stranded structure comes about through a process known as *base pairing*, in which the nitrogenous bases from one strand line up and hydrogen bond with the bases from another strand. The process is called *base pairing* because the nucleotide structures are such that each nucleotide only hydrogen-bonds in a stable way with a certain other nucleotide, its pair nucleotide. Guanine pairs only with cytosine. In DNA, adenine pairs with only thymine. In RNA, adenine pairs with uracil. Figure 7-20 shows hydrogen bonding between base pairs.

The helical secondary structure is due to base stacking interactions, rather than hydrogen bonding. This is true for both single-stranded and double-stranded nucleic acids. The double-stranded form, however, requires hydrogen bonding between the bases of each strand in order to form a stable double helix. And hydrogen bonding between nucleotide bases requires the specific base pairings as mentioned. Therefore stable double-stranded nucleic acids can occur only if the

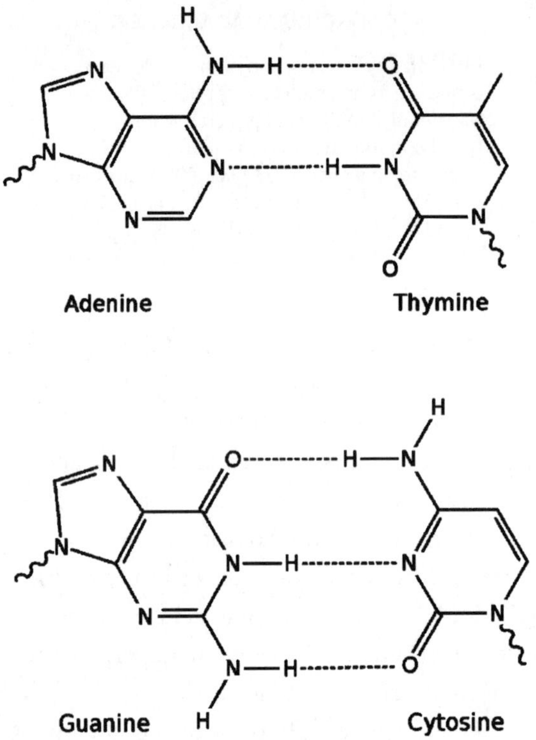

Adenine   **Thymine**

Guanine   H   **Cytosine**

FIGURE 7-20 • Hydrogen bonding between base pairs in DNA.

two nucleotide strands are *complementary*. Two nucleotide strands are *complementary* if, in every place on one strand where there is an adenine, the other strand contains thymine (or uracil) and vice versa. And in every place on one strand where there is a guanine, the other strand contains cytosine (and vice versa). Figure 7-21 shows hydrogen bonding between two strands of DNA.

Nucleic Acid Function

Nucleic acids serve a variety of functions in living things. Here we review only the most major roles they play. We start with DNA, whose primary role is *storage*, *expression*, and *replication* of genetic information.

We said that proteins are the worker bees of life; they play a major role in pretty much all biological process. This includes everything from digestion, to oxygen transport, to muscle contraction, sensory perception, eye color, hair color, and the manufacture of biomolecules such as carbohydrates, lipids, nucleic acids, and even proteins themselves, just to name some of the more major roles proteins play. You also learned (in Chap. 2) that the specific shape and physical

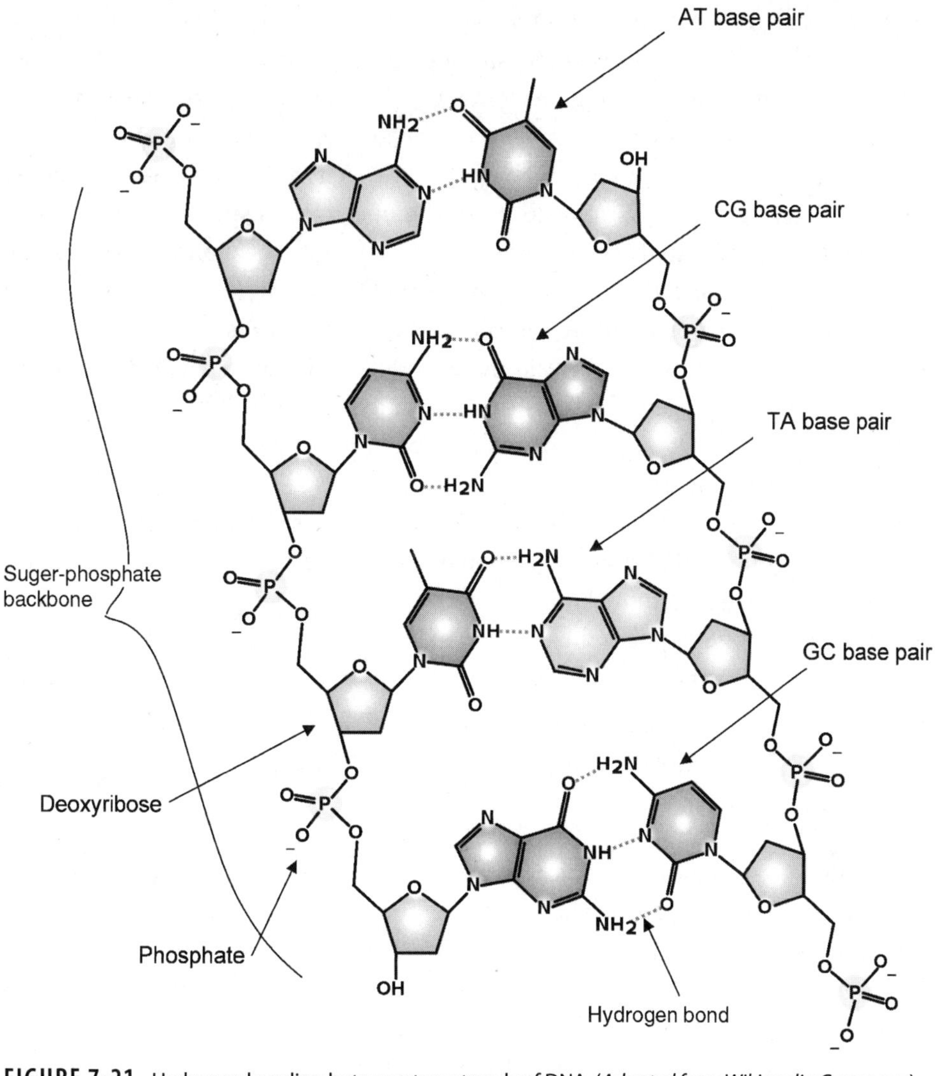

FIGURE 7-21 • Hydrogen bonding between two strands of DNA. (*Adapted from Wikimedia Commons.*)

properties of a protein are intrinsically related to its function. Finally, you learned a little earlier in this chapter that the specific shape and physical properties of a protein depend largely on two things: (1) its amino acid content (which amino acids make up the protein) and (2) the specific order or sequence of the amino acids in the polypeptide.

Given the major roles that proteins play, the enormous number of different proteins, and the fundamental connection between a protein's function and its amino acid content and sequence, how does an organism know which amino acids to use for each different kind of protein and in what order to put those amino acids together? This is where DNA comes in. The sequence of nucleotides in DNA is actually a *genetic code* specifying the content and sequence of amino acids in each protein. This is what is meant by *storage* of genetic information. Various sections of the nucleotide sequence along a DNA polymer specify the amino acid sequence for a particular protein. (Some sections of the nucleotide sequence, however, do not *appear* to specify anything; possible reasons for this will be mentioned later, but a complete discussion of these *noncoding regions* of DNA is more apt for an advanced text.) Each such section of DNA that specifies the amino acids for a given protein is called a *gene*. There is a concept in molecular biology of one gene, one protein. The idea is that each gene contains the instructions for a particular protein, and each different kind of protein is coded for by a particular gene.

In general one gene, one protein is true. But there is also a wide variety of exceptions to this rule. The actual situation is known to be quite complex. For example, sometimes a gene codes for only a portion of a protein. And sometimes parts of an amino acid sequence (although coded for by the gene) are removed by the organism before the polypeptide is finally assembled into a protein. Despite these complexities, for our purpose of gaining a basic understanding the biophysics of these molecules, the one gene, one protein rule will suffice. Figure 7-22 summarizes the role of DNA as a storehouse of genetic information.

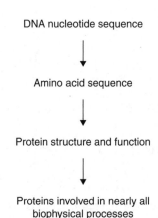

DNA nucleotide sequence

Amino acid sequence

Protein structure and function

Proteins involved in nearly all
biophysical processes

FIGURE 7-22 • DNA as a storehouse of genetic information. Genetic information (physical heredity) affects the entire organism through proteins. In other words DNA contains the blueprints for building proteins. Proteins, in turn, are involved in all living functions, including building all biomolecules. In this way, DNA is effectively the blueprint for the entire organism because it affects the entire organism by specifying how to build each protein.

One of the roles that DNA plays is *expression* of genetic information. By *expression* we mean actually reading the instructions stored in DNA's nucleotide sequence and starting the process of building a protein from them. To explain how this happens, we first tell the short story, then tell it again but with some details. Even so, what follows is just a basic description of gene expression; we'll leave the finer points for a text in molecular biology.

The short story is this: The DNA double helix unwinds to expose the nucleotide bases. A copy of the DNA nucleotide sequence is then made. This copy is in the form of a single-stranded ribonucleic acid (RNA). The single-stranded RNA molecule then moves to where protein synthesis will take place in the cell. We call this type of RNA *messenger RNA* (*mRNA*) because, like a messenger, it carries the blueprints from the DNA to the place where they are used to build a protein. Next, a complex group of proteins and RNA molecules read the nucleotide sequence in the mRNA and step by step link together the correct amino acids in the correct order (according to the instructions in the mRNA) to make a protein.

That's the short story; now a little more detail: When the DNA double helix unwinds at the start of a gene (the beginning of a sequence that codes for a particular protein), there are individual *ribonucleotides* floating around in solution. These ribonucleotides form hydrogen bonds with the deoxyribonucleotides on one of the exposed strands of the DNA. Keep in mind that hydrogen bonding between nucleotide bases is pair specific. Then an enzyme called *RNA polymerase* catalyzes the biochemical reaction to covalently link the ribonucleotides together. The result is the single-stranded mRNA molecule whose base sequence is now *complementary* to the DNA sequence. Effectively the RNA molecule is a copy or *transcript* of the base sequence in the gene. So the process of synthesizing mRNA from DNA is called *transcription*. See Fig. 7-23.

When protein synthesis takes place, a number of enzymes and ribonucleic acids get involved. One major player is a type of RNA called *transfer RNA*, (*tRNA*), whose job it is to transfer each amino acid to the correct location in the growing polypeptide chain. There are a number of different types of tRNA, one for each amino acid. Each tRNA molecule has two specific binding sites. One part of the tRNA molecule binds to a specific amino acid. The structure of this part of the tRNA molecule and the structure of the amino acid are such that the tRNA molecule binds specifically to that amino acid but not to others. The second binding site of the tRNA molecule has a specific nucleotide sequence, three nucleotides long, that can base-pair with the complementary sequence on the mRNA. In this way the tRNA molecules *translate* the mRNA sequence into a sequence of amino acids. Each three nucleotides in the mRNA are translated into a specific

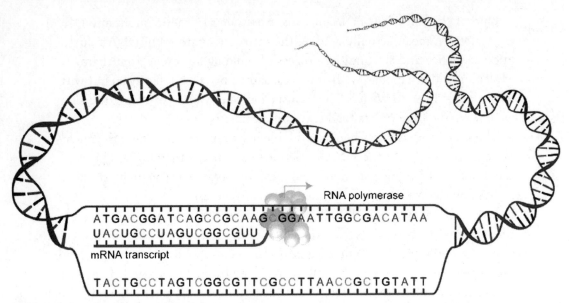

RNA polymerase

ATGACGGATCAGCCGCAAG GGAATTGGCGACATAA
UACUGCCUAGUCGGCGUU

mRNA transcript

TACTGCCTAGTCGGCGTTCGCCTTAACCGCTGTATT

FIGURE 7-23 • Transcription of DNA to create an mRNA transcript, or copy, of the nucleotide sequence in a gene. (*Courtesy of National Human Genome Research Institute.*)

amino acid. Figure 7-24 illustrates the process of *translation*, translating the sequence of nucleotides in mRNA into a sequence of amino acids in a polypeptide chain.

Related to the idea of *genetic expression* is the concept of *gene regulation*, how the organism decides when to express a particular gene versus when it's time to "turn the gene off" and stop building the particular protein. The organism uses a wide variety of mechanisms for gene regulation. Without going into detail at this time, it's worth mentioning some general themes involved in gene regulation. These include proteins binding to DNA (to encourage or discourage unwinding of the double helix), various other molecules binding to the proteins involved in gene expression, and conformational changes in proteins and/or DNA.

Now that we know where the sequence of amino acids in a protein comes from, we could ask where the nucleotide sequence in DNA comes from. This is where inheritance and *DNA replication* come in. One of the roles of DNA is to *replicate* itself so that the nucleotide sequence can be passed on to the next generation. The process of *DNA replication* has some similarities with transcription. The double helix unwinds to expose the nucleotides, but now *deoxyribonucleotides* in solution pair with the deoxyribonucleotides in the DNA strands. The enzyme *DNA polymerase* then does the work of covalently connecting together these nucleotides into a new DNA strand.

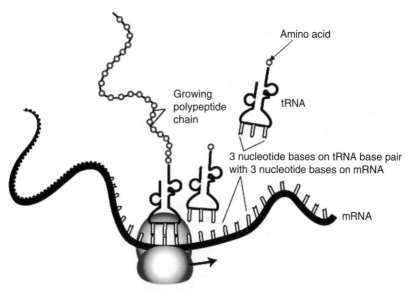

FIGURE 7-24 • Protein synthesis (building a protein from amino acids) involves translation of an mRNA nucleotide sequence into an amino acid sequence by tRNA molecules.

DNA replication, in contrast to transcription, uses *both* DNA strands as a template to create a new nucleotide polymer. Figure 7-25 illustrates the process of DNA replication. The newly formed DNA strands do not separate from the original DNA. Instead, the replication results in *two* new DNA double helices, where each new double helix is a combination of one strand from the original DNA molecule and a newly synthesized complementary strand. Once the DNA has been replicated, a copy of the DNA can be passed on to the next generation through cellular division (one cell dividing into two).

Nucleotides as Energy Currency

One of the most important roles of nucleotides (outside of being the residues that make up nucleic acids) is that of being energy currency for the cell. As we mentioned, nucleotides sometimes contain two or even three phosphate groups in a row. (See, e.g., Fig. 7-26.)

The covalent bonds between adjacent phosphate groups (*pyrophosphate bonds*) contain a relatively large amount of energy. This energy is readily available by breaking these bonds. Many enzymes couple simultaneous reactions that include the breaking of a nucleotide pyrophosphate bond with a reaction that requires energy. In this way the energy of the pyrophosphate bond is used to drive biophysical processes that would otherwise not be spontaneous. The nucleotide is

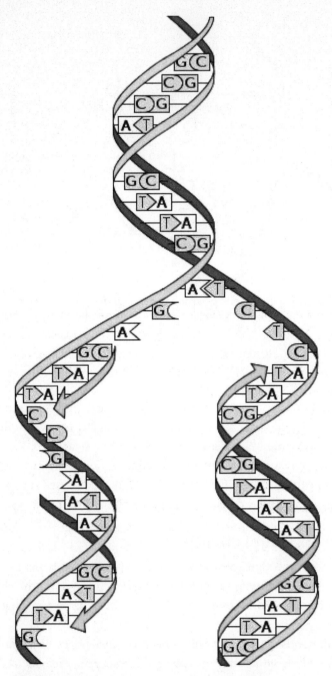

FIGURE 7-25 · DNA replication. Each new double helix consists of one strand from the original DNA molecule, plus a newly synthesized complementary strand. (*Courtesy of Wikimedia Commons.*)

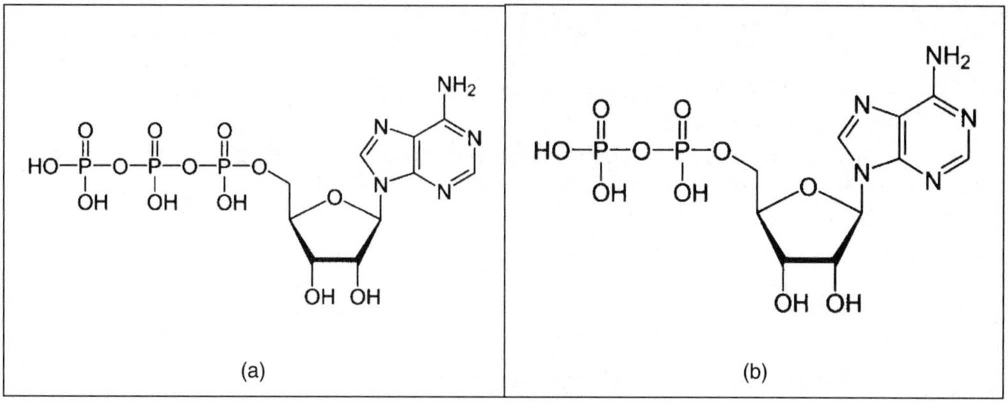

FIGURE 7-26 • Adenosine triphosphate (ATP) and adenosine diphosphate (ADP).

acting as energy currency, its only role in the process being to provide energy where it is needed.

As an example, in the manufacture of proteins, attachment of an amino acid to its tRNA requires converting one molecule of ATP (adenosine triphosphate) to ADP (adenosine diphosophate). The energy released by breaking off the third phosphate group (converting ATP to ADP) is used to form the chemical bond between the amino acid and its tRNA. When the amino acid is then added to the growing polypeptide chain, two GTP (guanosine triphosphate) molecules are converted to GDP (guanosine diphosophate) to provide the energy needed to form the peptide bond.

By far the most common nucleotide used as an energy currency is adenosine triphosphate (ATP). We mentioned regarding lipids and carbohydrates that among their roles is the role of storing energy. What's the difference between storing energy in lipids and carbohydrates versus the energy stored in the pyrophosphate bonds of ATP? The answer is that the energy in ATP is readily used in many biophysical processes, whereas the energy in carbohydrates and lipids usually must first be put into pyrophosphate bonds before it can be used further to drive biochemical reactions. You can think of the energy in carbohydrates and lipids as money in the bank whereas ATP is like cash in your hand. When the organism needs more energy, it breaks down the carbohydrates and lipids, using the energy in these biomolecules to convert ADP back into ATP. The ATP then can be used to drive other biophysical processes such as manufacture of proteins and nucleic acids, pumping of ions across membranes, and so on.

QUIZ

Refer to the text in this chapter if necessary. Answers are in the back of the book.

1. The process of converting the nucleotide sequence of an mRNA molecule into an amino acid sequence in a polypeptide is called what?
   A. Transcription
   B. Translation
   C. Transformation
   D. Trans fat

2. Carbohydrates contain what three elements?
   A. Carbon, hydrogen, and rategen
   B. Nitrogen, oxygen, and carbon
   C. Oxygen, carbon, and hydrogen
   D. CHNOPS

3. A major role of lipids is what?
   A. Muscle contraction
   B. Phosphorylation of DNA
   C. Structure of phospholipids
   D. Structure of membranes

4. Carbon atoms usually form how many covalent bonds?
   A. 2
   B. 4
   C. 6
   D. 8

5. An amino acid is a molecule with what functional groups?
   A. An amine and hydrochloric acid.
   B. A hydroxyl and an amine.
   C. A protein and a polypeptide.
   D. A carboxyl and an amine.

6. DNA and RNA are examples of
   A. nucleic acids.
   B. nucleotides.
   C. proteins.
   D. lipids.

7. DNA is a
   A. heteropolymer.
   B. homopolymer.

    C. peptide polymer.

    D. ribonucleotide polymer.

8. **The helical shape of DNA is due primarily to**

    A. aromatic side chains.

    B. base stacking.

    C. hydrogen bonds.

    D. van der Waals constant.

9. **Proteins are involved in**

    A. DNA synthesis.

    B. enzymatic processes.

    C. lipid metabolism.

    D. all of the above

10. **Which of the following is *not* true about proteins?**

    A. Proteins are polymers of amino acids.

    B. Polypeptide chains fold up into a specific shape necessary for the protein's function.

    C. Transfer RNAs carry amino acids to where they are needed for protein production.

    D. Peptide bonds hold base pairs together.

chapter 8

The Cell

We want to dig deeper into the physics of biomolecules. In order to do so, we must first understand the context in which biomolecules function. That context is almost always the cell. We also want to lay a foundation to study the physics of the cell itself. With this in mind, in this chapter we review the main structures of living cells and give you a basic understanding of the relationship between cell structures and the biomolecules they are made of.

CHAPTER OBJECTIVES

In this chapter, you will

- Learn the basic concepts of cell theory.
- Learn the structures of the cell and their functions.
- Learn the relationship between biomolecules and cell structures.

What Is a Cell?

So what is a cell? A *cell* is a pouch or sack of biomolecules *that is* (or was) *alive.* A cell is the smallest unit of an organism that can be considered alive. If you break the cell into anything smaller, it's just a bunch of biomolecules and not considered alive. The pouch or sack is part of the cell itself and is called the *cell membrane.*

So what do we consider alive? Something is *alive* if it *can take in matter and energy and use that matter and energy to do work on its surroundings, to grow, and to reproduce itself.*

This definition leaves out certain things, about which scientists are still debating whether they should be considered organisms. For example, a virus is a DNA molecule surrounded by a protein coat. Should it be considered alive? Viruses reproduce, but they do not grow. And viruses only reproduce with the aid of a cell they've infected. Without a host cell a virus can't do much of anything. So for now, for our purposes, we will consider living things to be only those things that meet our definition's criteria concerning matter and energy and growing and reproducing. (*Note:* This definition of *alive* is not meant to exclude individual organisms or individual cells that may be at some phase of life lacking in one of these features. E.g., an adult human being stops growing in terms of their body height, but their cells are still growing and reproducing, so the person is alive. Or, e.g., an individual cell may stop reproducing, while other cells in the same organism are still reproducing and the individual cell that has stopped is still otherwise functioning—taking in matter and energy to do work—and so that individual cell is alive.)

Cell theory was formally developed in 1839 by M. J. Schleiden and T. Schwann. The theory was based not only on Schleiden's and Schwann's own microscopic observations, but also the observations and ideas expressed by others going as far back as the mid- to late-1600s, when scientists including Robert Hooke and Antonie van Leeuwenhoek began using microscopes to examine living things.

Cell theory says (1) all living things are made of cells and (2) all cells come from preexisting cells.

A quick note regarding the second point: many people have tried unsuccessfully to create a living cell from biomolecules. As of this writing, it is still the case that all living cells we know of came from previously existing cells. The process by which cells come from preexisting cells is called *cell division.* In cell division, a cell replicates its DNA and then splits into two daughter cells. Each of the daughter cells gets a copy of the replicated DNA.

An organism can be *single cellular*, that is, made up of a single cell (like an amoeba) or *multicellular* (like people). Cells themselves can grow, by taking in matter and energy to build up their internal structures and increase their membrane size. Organisms can grow too (obviously). Single-cell organisms grow as a cell grows. Multicellular organisms grow primarily by adding cells to themselves through cell division.

Cells can be large or small. Most cells are too small to see with the naked eye. However, a modest microscope is usually sufficient to visualize most kinds of cells. In some cases cells can be quite large. For example, an egg is a single cell. The eggs of many organisms are too small to see without a microscope, whereas the eggs of others are big enough to eat for breakfast.

There are many different types of cells. The cells of different organisms can be very different from one another. But even within the same organism cells specialize to carry out different functions. For example, nerve cells are long and thin like wires to carry nerve impulses throughout the body. Muscle cells can contract to provide movement. Skin cells are flat and strong to protect the body. And bone marrow cells manufacture blood cells which carry oxygen through the blood. The process of a young cell maturing to specialize and become a particular type of cell is known as *differentiation*. Figure 8-1 shows some examples of various types of cells. For the most part, however, in the remainder of this chapter we will concern ourselves mainly with those things that cells have in common.

Cell Structure

All cells are surrounded by a lipid bilayer *membrane* that encloses and contains the contents of the cell. We already learned that many lipids are amphipathic, having both hydrophobic and hydrophilic parts on the same molecule. In an aqueous or polar environment these molecules arrange themselves in complexes (aggregates of molecules) that place their hydrophobic parts inside, away from the water, and their hydrophilic parts on the outside, facing the water. One such complex is a lipid bilayer as shown in Fig. 8-2. The bilayer forms a membrane around the cell. The polar, hydrophilic portions of lipid molecules face the aqueous environments on both the inside and outside of the cell. The nonpolar, hydrophobic parts lie in the center of the bilayer and form a barrier that blocks many molecules from crossing the membrane. This keeps the inside of the cell on the inside and the outside on the outside. The cell membrane is in fact *selectively permeable*, meaning that certain types of molecules can pass through the membrane, whereas others cannot.

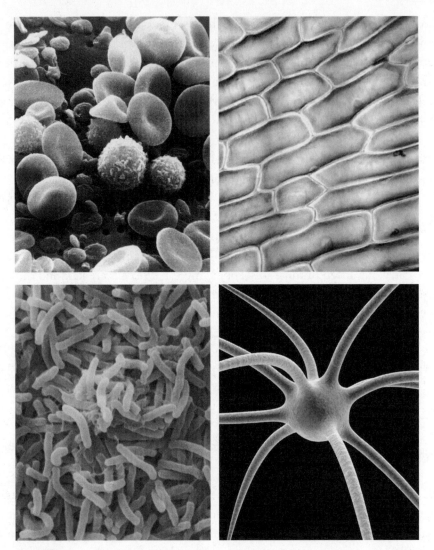

FIGURE 8-1 • Microscopic images of various cell types, clockwise from top left: blood, onion, nerve, and cholera bacteria. (*Courtesy of Wikimedia Commons.*)

The inside of the cell is called the *cytoplasm*. It contains all the biomolecules we have discussed so far, and more. This is where these molecules do the work of keeping the cell alive.

There are two major categories of cells, and all organisms are made up entirely of one of these two categories of cells. Some cells contain a central, membrane-bound compartment called the *nucleus* which contains most or all of the cell's genetic material. Cells that have a nucleus are called *eukaryotic*

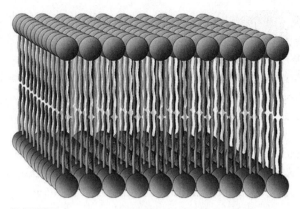

FIGURE 8-2 · A lipid bilayer, the foundation of cell membranes, separates the *cytoplasm* (the inside of the cell) from the extracellular environment.

cells, and the organisms that contain these cells are called *eukaryotes*. Cells that lack a nucleus are called *prokaryotic* cells, and organisms made of *prokaryotic* cells are called *prokaryotes*.

There are some important distinctions between eukaryotes and prokaryotes. Eukaryotes, in addition to a nucleus, have other membrane-bound, internal structures called *organelles*. These organelles *compartmentalize* some of the functions of the cell. For example, an organelle called a *mitochondrion* carries out the role of generating ATP molecules for use as energy currency throughout the cell (see Chap. 7). Some of these organelles are shown in Fig. 8-3, which highlights the structures of prokaryotic and eukaryotic cells.

In contrast to a eukaryotic cell, a prokaryotic cell is one large compartment. All of the cell's functions are carried out in this one compartment. However, like a one-room apartment, various functions tend toward certain locations

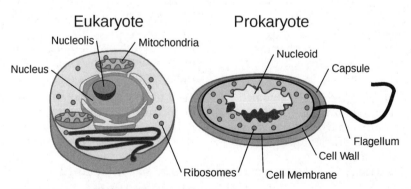

FIGURE 8-3 · Eukaryotic versus prokaryotic cells.

within the cell. For example, in prokaryotes, protein synthesis occurs for the most part near the edge of the cytoplasm, just inside the cell membrane. And DNA and its functions are found mostly toward the center of the cell.

There are two main divisions of prokaryotes: bacteria and archaea. Prokaryotes were originally thought of as unicellular plants, because of similarities they have with plant cells. However, with the advent of detailed genetic analysis (analysis of the organism's nucleic acid sequence), it now makes sense to classify bacteria and archaea into domains of their own, separate and apart from plants and animals. Prokaryotic organisms are almost exclusively unicellular; however, many can form multicellular colonies that have properties resembling multicellular organisms. Eukaryotes are for the most part multicellular organisms, such as plants and animals. But unicellular eukaryotes exist as well (e.g., the amoeba).

Table 8-1 summarizes some of the differences between eukaryotes and prokaryotes.

TABLE 8-1   Eukaryotes and prokaryotes compared.

Eukaryotes	Prokaryotes
Nucleus and other membrane–bound organelles compartmentalize the cell.	No nucleus. Genetic material and all biomolecules are all together in the cytoplasm.
Mostly multicellular; some unicellular.	Unicellular (some have colonial or multicellular stage of life).
Slower division rates.	Rapid division rates. Can adapt quickly to drastic changes in environment.
Can carry out more complex and more efficient biochemical reactions, because reactions can be isolated in membrane–bound compartments where reactants can be brought closer together.	Biochemical reactions are simpler, and must all occur in the cytoplasm where more complex reactions might interfere with each other. Also, less efficient in bringing reactants together.
Tend to be larger, since able to ingest nutrients actively and process those nutrients with a wide variety of reactions for efficient growth and maintenance. Typical size is 10 to 100 µm across.	Tend to be smaller. Nutrients obtained inefficiently, primarily through diffusion, and processed with only a simple set of biochemical reactions. Typical size is 0.1 to 10 µm across.
Plants and animals.	Bacteria and archaea.

The Cell Membrane

As mentioned, all cells are surrounded by a lipid membrane, which keeps the insides in and the outsides out. The cell membrane is also called the *plasma membrane*. In addition to lipids, cell membranes contain a variety of molecules embedded in, or connected to, the lipid bilayer. These include proteins, carbohydrates, glycoproteins (proteins with carbohydrate attached), and glycolipids (lipids with carbohydrate attached). These molecules are often classified in terms of how deep they are within the membrane. If such a molecule is embedded deep into the membrane, we call it *integral*, for example, integral membrane protein. Those that are just on the surface (but are still fundamentally part of the membrane) are called *peripheral*. If a molecule is embedded all the way through the membrane, sticking out on both sides, it is called *transmembrane*. Figure 8-4 illustrates these, and some other, features of cell membranes.

Membrane Transport

Both the inside and outside of the cell are typically aqueous environments. In order for a molecule to dissolve in an aqueous environment such as blood or cytoplasm, the molecule must be polar or hydrophilic. But for a molecule to penetrate the membrane's lipid barrier, the molecule must be lipid soluble, that is, nonpolar or hydrophobic. So the hydrophobic portion of the membrane acts as a barrier to keep molecules from passing through. Some small molecules are able to pass through the membrane, for example. water or carbon dioxide, because their interactions with the lipids are minimal.

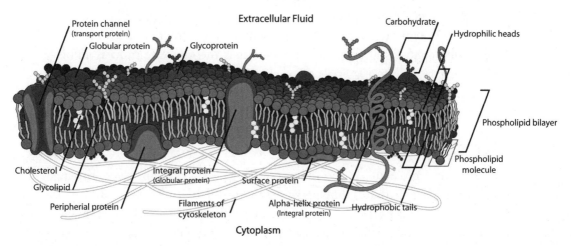

FIGURE 8-4 • Cell membrane.

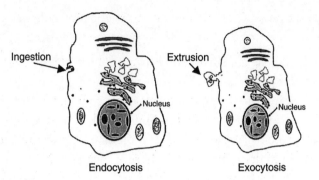

FIGURE 8-5 • Endocytosis and exocytosis. (*Courtesy of* Biochemistry *Demystified.*)

Although the membrane acts to protect the cell from larger outside molecules, intruders, or environmental hazards, the cell does need to somehow get nutrient molecules in and waste molecules out. The cell has a variety of ways to *transport* molecules through the membrane when it needs to. The physics of membrane transport is discussed in Chap. 11. Here we outline the various transport methods, and some of the molecules involved.

One method is via *membrane transport proteins*. These are *transmembrane* proteins that assist other molecules in getting through the membrane. They may simply provide a channel or tunnel for the molecule to pass through. Or they may bind the other molecule and then undergo a conformational change that essentially pulls the other molecule through the membrane.

Two other methods are *endocytosis* and *exocytosis*. See Fig. 8-5. In endocytosis, a portion of the cell membrane folds in toward the cell. At the same time, this portion of membrane engulfs some molecule or molecules from outside the cell. This portion of membrane then breaks off from the membrane and carries the engulfed contents into the cell. Exocytosis is the opposite process and is used by cells to secrete materials or to eliminate waste. In exocytosis a small membrane pouch of material merges with the cell membrane, releasing its contents outside the cell.

Receptors

A receptor is a molecule in the membrane (such as a protein, glycoprotein, glycolipid, or carbohydrate) that has at least a portion of itself on the surface of the membrane. This portion of the molecule on the surface of the membrane has *specificity* for some other molecule. This means that only one particular type of molecule can bind to the receptor. For example, a particular receptor may bind a specific amino acid. The binding of a specific molecule to its receptor is often a *signal* for the cell to do something. For example, binding of a certain molecule to a receptor might signal to the cell that it should increase its production of a certain enzyme. It might be a signal to the cell to begin endocytosis of some foreign object; this is the case for example when a white blood cell engulfs and destroys a virus particle.

Sometimes receptors are also transport proteins. If the receptor is also a transport protein, it may then pull that bound molecule through the cell membrane into the cell. Sometimes receptors signal the need to transport another molecule. For example, the hormone insulin (a small protein) binds to insulin-specific receptors in the cell membrane. This signals glucose transport proteins elsewhere in the membrane to allow glucose into the cell.

Cell Walls

Some cells have a protective layer on the outside surface of the membrane called a *cell wall*. Cell walls are commonly found in plants and many prokaryotes such as bacteria, but are not usually found in animal cells. The cell wall is a relatively stiff structure, enabling the cell to maintain a specific shape even under a certain amount of external or internal pressure.

In plants the cell wall is commonly made from cellulose, a polymerized carbohydrate. In other organisms cell walls may be made from glycoproteins or a variety of polysaccharides. Although cell walls are somewhat flexible, they are more rigid than the cell membrane. This additional rigidity means that organisms with cell walls are usually not capable of endocytosis and exocytosis. Instead they rely heavily on membrane transport proteins. Some bacteria, called *gram-negative bacteria*, have an extra membrane on the outside surface of the cell wall. (The term *gram-negative* comes from the fact that this extra membrane prevents the cell from absorbing Gram stain. Gram staining is a technique for highlighting and visualizing certain differences among cells and the structures inside them. The technique was developed by H. C. Gram in 1882.)

Cell Organelles

We now list and describe the major internal structures, or *organelles*, found inside a cell. For the most part, with the exception of the cell nucleus, there may be several or many instances of each of these organelles. Also, the following list is not exhaustive, but is just intended to introduce you to the more important organelles of the cell.

You should keep in mind, however, that these structures are not, by any means, the only stuff inside the cell. In addition to the organelles listed below, the inside of the cell is a densely populated environment with hundreds, if not thousands, of different *types* of molecules. These include water, dissolved gasses such as oxygen and carbon dioxide, structural proteins, enzymes, nucleic acids, lipids, various ions (sodium, calcium, magnesium, potassium, chloride, phosphate, etc.), minerals, vitamins, sugars and other nutrients and all the building blocks of these molecules, for example, nucleotides, small polypeptides, amino acids, and a whole host of small organic molecules that are part of the thousands of biochemical reactions and biophysical processes continually happening inside the cell.

The amount of each type of molecule inside a cell varies widely, anywhere from a few copies for some molecules to hundreds of millions for others. Altogether, depending on the size and type of cell, a typical single cell may be made up of anywhere from hundreds of billions to tens of trillions of molecules.

Taken together, everything inside the cell (other than the cell membrane) is called the *cytoplasm;* this includes all of the structures listed in the following sections, plus all of the molecules floating around inside the cell (some of which were just listed).

DNA

DNA is perhaps the largest and certainly the most significant of structures inside the cell. The cell's DNA is its genetic material, the instructions passed down from generation to generation for building all of the proteins in the cell, which in turn do the work of building, mediating or getting involved in pretty much everything else in the cell. The DNA in the cell is organized into structures called *chromosomes*. Each *chromosome* is a single DNA molecule that has been packaged and organized in a specific way. Sometimes the DNA is organized into complexes with other molecules such as proteins. This is typically the case with most DNA in eukaryotes. In prokaryotes the DNA may be packaged by coiling it up tightly.

At certain stages of a cell's life cycle, the DNA is more spread out, like a plate of spaghetti, and individual chromosome structures are not apparent. In this state, the DNA is referred to as *chromatin*. See Fig. 8-6. All of a cell's chromosomes, or all of a cell's chromatin, taken together as a whole is called the cell's *genome*.

Prokaryotes almost always contain only one chromosome, a single DNA molecule. That's the entire prokaryotic genome, just one chromosome. In prokaryotes, the ends of this single DNA molecule are covalently connected to one

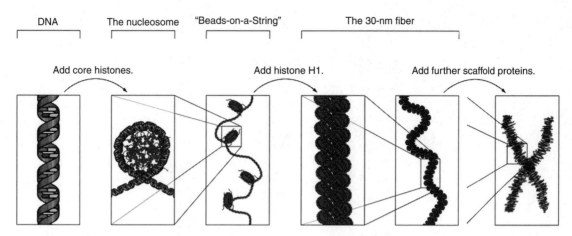

DNA   The nucleosome   "Beads-on-a-String"   The 30-nm fiber

Add core histones.   Add histone H1.   Add further scaffold proteins.

FIGURE 8-6 · Structure of eukaryotic chromatin. A *nucleosome* is a DNA double helix wrapped around a cluster of proteins called a *histone*. Since the DNA helix is itself wrapped into the shape of a coil or helix, this configuration is called a *supercoil* or *superhelix*. Nucleosomes are then further packaged together with protein histone H1 to create 30-nm fibers of chromatin. These are further packaged with other scaffolding proteins to make a chromosome structure. (*Figure derived from Wikimedia Commons.*)

another, making the chromosome one big DNA circle. This DNA circle is usually so long and thin that, although it's a circle, it still resembles a plate of spaghetti.

The Cytoskeleton

Cells contain many fibrous structures which together are called the *cytoskeleton*. These tubelike and ropelike structures are often interconnected, and play a number of roles in which they support, transmit, or apply forces, in much the same way that bones support, transmit, and apply forces for the body (thus the term *skeleton*). Among other things, the cytoskeleton helps to support and maintain a cell's shape, it transmits and applies forces for cellular movement, and it helps anchor various organelles into place within the cell. Both eukaryotes and prokaryotes have a cytoskeleton.

Cytoskeleton fibers are made of polymerized protein molecules. (Remember, proteins are themselves polymers of amino acids, so the cytoskeleton fibers are polymers where each residue is itself a folded up polymer.) There are three types of cytoskeleton fibers, classified according to their size and shape: *microtubules*, *intermediate filaments*, and *microfilaments*.

Microtubules are the largest, about 25 nm in diameter. They are long tubelike structures made up from the protein tubulin. There are two types of tubulin: α-tubulin and β-tubulin. One α-tubulin molecule and one β-tubulin molecule combine to form a *dimer* (a polymer of two molecules). The tubulin dimers then

FIGURE 8-7 · Illustration of cytoskeleton fibers, from left to right: microtubules, intermediate filaments, and microfilaments.

polymerize end to end, with the α-tubulin from one dimer connecting to the β-tubulin from the next. This creates what is called a *protofilament*. The protofilaments then bundle together, to form a strong tubelike structure. See Fig. 8-7. Intermediate filaments and microfilaments are also polymers of proteins, just different proteins, for example, the proteins actin or keratin. Also, as their names imply, intermediate and microfilaments are not *tube*like. Instead their protofilaments combine to form structures similar to a twisted cable or rope.

As mentioned, one of the roles of the cytoskeleton is in movement. The cytoskeleton is involved both in moving things around inside the cell and in moving the cell itself. This is a fascinating area of biophysics. The cytoskeleton generates forces for movement through the polymerization (or depolymerization) process. As residues are added to or removed from one end of a cytoskeleton fiber, then as the fiber grows or shrinks, it pushes or pulls whatever is attached to the other end. So, for example, if a part of the membrane is attached to the other end, the cytoskeleton can generate the forces needed to bend the membrane during endocytosis.

Ribosomes

Both eukaryotes and prokaryotes contain large complex structures called *ribosomes*. A *ribosome* is a complex of structural proteins, enzymes, and ribosomal RNA (rRNA). Ribosomes are factories where protein synthesis takes place. (See Chap. 7, the section Nucleic Acid Function, for a description of protein synthesis.) The ribosome is the

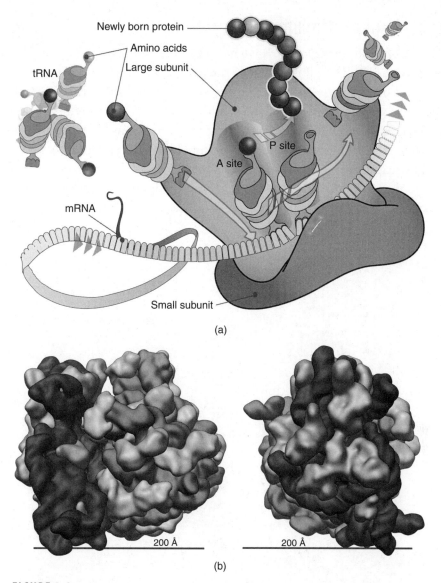

FIGURE 8-8 · (a) Illustration of ribosome and its role in protein synthesis. (b) Computer-generated front and side view of a ribosome, from the bacteria, Escherichia coli. (*Courtesy of Wikimedia Commons.*)

place where the messenger RNA, the transfer RNA, and the amino acid all get together. The ribosome ensures that these three things work together in just the right way needed to polymerize the amino acids according to the sequence specified by the mRNA. See Fig. 8-8 for illustration. There is evidence that some of the catalytic (enzymatic) activity needed to synthesize proteins comes not only from the ribosomal proteins but also from the ribosomal RNA.

Endoplasmic Reticulum

The *endoplasmic reticulum* (*ER*) is a network of folded membranes found inside eukaryotic cells. The ER is similar to the plasma membrane in that it is a phospholipid bilayer with various other molecules embedded within it. However, the ER is a *folded membrane* that doesn't actually surround anything. Rather it provides a network of membrane surfaces and pathways between them. See Fig. 8-9. In general in cells, wherever we find folded membranes, we find many biophysical and biochemical processes associated with the membrane. The folding of the membrane creates a much larger membrane surface (for a given amount of volume) to facilitate these processes.

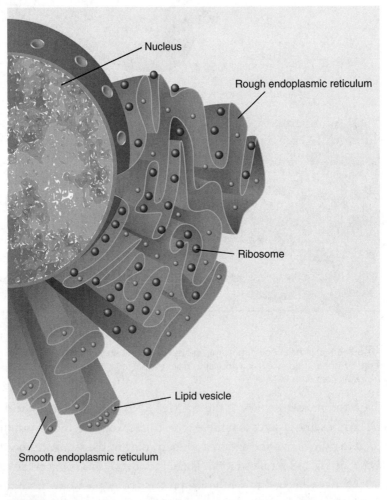

FIGURE 8-9 • Endoplasmic reticulum. (*Derived from Wikimedia Commons.*)

There are two types of endoplasmic reticulum, smooth and rough; the main difference is that the rough ER has ribosomes on its surface. The rough ER is involved in protein synthesis, whereas the smooth ER is involved in lipid and steroid synthesis. Both types of ER are also involved to some extent in additional processing of molecules, for example, adding carbohydrates to proteins to form glycoproteins, splicing and folding of polypeptide chains as part of protein formation, and packaging of proteins and other molecules into lipid vesicles for transport to other parts of the cell.

Nucleus

The *nucleus* is an approximately spherical membrane-bound (membrane-surrounded) organelle near the center of *eukaryotic* cells. The nucleus contains almost all of the cell's genome. (Some genetic material is also found in mitochondria.) The main functions of the nucleus are gene expression (transcription of genes into RNA to make proteins) and DNA replication (copying of the genome just prior to cell division). The nucleus is often referred to as the control center of the cell. This is because the nucleus controls the expression of almost all of the cell's genes, which in turn control the manufacture of the cell's proteins, which in turn are involved in virtually every biochemical and biophysical process inside the cell.

The nucleus is surrounded by a double membrane called the *nuclear envelope* or *nuclear membrane*. By *double membrane*, we mean that it consists of *two* membranes, each of which is itself a lipid bilayer (with various other molecules embedded in it). The space between the inner nuclear membrane and the outer nuclear membrane is called the *perinuclear space*. The outer nuclear membrane is contiguous (directly connected to) to the membrane of rough endoplasmic reticulum. This connection helps facilitate protein synthesis which occurs on the rough ER since it is directly dependent on the transcription of genes into mRNA, tRNA, and rRNA in the nucleus. Along the nuclear membrane are special proteins called *nuclear pores*. Nuclear pores are membrane transport proteins that span *both* membranes of the nucleus and control the flow of molecules into and out of the nucleus. See Fig. 8-10. Just inside the inner nuclear membrane is the *nuclear lamina*, a network of intermediate filaments that plays a role similar to the cytoskeleton. The nuclear lamina provides mechanical support, aids in transport of molecules to and from nuclear pores, and is involved in organizing chromatin. It also helps separate chromosome copies during cell division to ensure that each daughter cell gets its own copy of each chromosome.

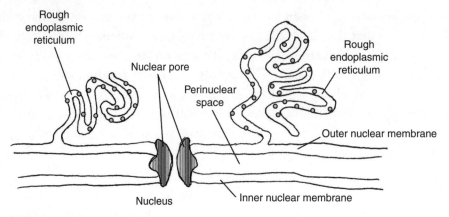

FIGURE 8-10 · Nuclear envelope.

Endomembrane System

In addition to the endoplasmic reticulum, there are a number of membranous organelles that work together in eukaryotic cells. Together, these organelles are called the *endomembrane system*. Figure 8-11 illustrates the endomembrane system.

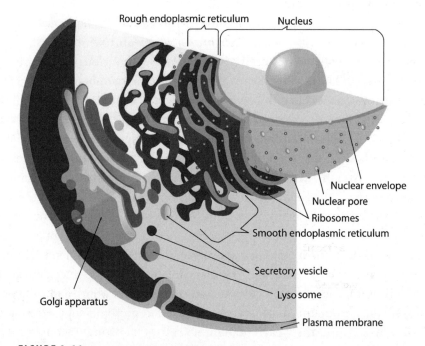

FIGURE 8-11 · Endomembrane system.

Golgi Apparatus

One of the larger organelles in the endomembrane system is the *Golgi apparatus*, also called the *Golgi complex* or *Golgi body*. The Golgi apparatus is very similar to smooth ER in that it is a folded membrane involved in processing and packaging of large molecules such as proteins and lipids. Many molecules manufactured and processed in the ER are later passed on to the Golgi apparatus for further processing and for packaging to be transported outside the cell. In addition, Golgi bodies are involved in the breakdown of carbohydrates and lipids so that the parts can be used in other processes inside the cell. The Golgi apparatus is named for Camillo Golgi who discovered it in 1898.

Vesicles

Vesicles are small, approximately spherical lipid bilayer containers. They typically result from a portion of some larger membrane structure budding out and being pinched off. In the process of budding out and being pinched off, various molecules (proteins, etc.) become trapped inside the vesicle. The contents of the vesicle can then be transported to other locations. Vesicles also commonly reverse the process, by fusing with other membrane structures and releasing their contents at the same time. When the vesicles fuse with or bud from the plasma membrane, these processes are called *endocytosis* and *exocytosis* (see earlier in this chapter). However, vesicle budding and vesicle fusion also occur frequently with other membranous structures such as the ER and Golgi apparatus.

Some types of vesicles are given specific names related to their contents and function. For example, *lysosomes* are vesicles that contain enzymes (called *lysozymes*) that specialize in breaking down or digesting larger molecules. In this way the somewhat harsh environment needed to digest larger molecules is isolated inside the lysosome to protect other molecules inside the cell. *Peroxisomes* are similar to lysosomes, but specialize in the breakdown of long chain fatty acids.

Vacuoles

Vacuoles are, in a sense, giant vesicles. They are often made from many vesicles that have fused together. Vacuoles have no particular shape. They are found in all plant cells and in some animal and bacteria cells. They serve a variety of functions, but are most commonly involved in isolating things that might otherwise be harmful for the cell and in containing and eliminating waste products (i.e., exporting waste from the cell via exocytosis). They are also involved in maintaining correct hydrostatic pressure inside the cell and using that pressure to support plant structures such as flowers and leaves.

Mitochondria

Mitochondria (singular *mitochondrion*) are membrane-bound organelles in most eukaryotic cells. Their main job is the synthesis of ATP (adenine triphosphate) which is used as the energy currency for many biophysical processes inside the cell. In this way mitochondria convert the energy stored in food into high-energy phosphate bonds of ATP. It is because of this role of mitochondria that they are sometimes called the *power plants* of the cell. Figure 8-12 illustrates the structure of a mitochondrion. There is an outer membrane and a folded inner membrane. The membrane folds that reach in toward the center of the mitochondrion are called *cristae*. The folded inner membrane supports the synthesis of ATP.

Mitochondria also contain DNA. The DNA molecules in mitochondria are small circular molecules. By small, we mean approximately 15 thousand base pairs long (as compared to about 1 million base pairs for a typical bacterial genome and about 3 billion base pairs in the human genome). There are typically 2 to 10 DNA molecules per mitochondrion. *Mitochondrial DNA (mtDNA)* replicates independently of the DNA in the nucleus. While the vast majority of proteins found in mitochondria are coded for by nuclear DNA, some of the proteins and RNA found in mitochondria are coded for by the mtDNA. For example, there are 37 genes in human mtDNA. Of these, 22 code for transfer RNA, 13 genes code for proteins, and 2 code for ribosomal RNA.

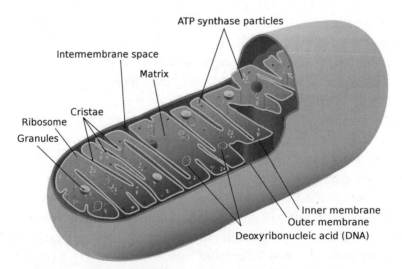

FIGURE 8-12 · Mitochondrion.

Chloroplasts

Chloroplasts are organelles found in the cells of plants and a few other eukaryotes. They are found mostly in the cells of leaves and other green portions of plants. Chloroplasts carry out the process of *photosynthesis*, which is the capture of energy from light and the conversion of that energy into the chemical bond energy of carbohydrates. This is how plants manufacture food from light and carbon dioxide. (Some bacteria also carry out photosynthesis, but they do so without chloroplasts which are only found in eukaryotes.)

Chloroplasts contain an outer membrane and an inner folded membrane. See Fig. 8-13.

The inner folded membrane forms a series of disklike or cylindrical structures called *thylakoids*. It is within the membranes of the thylakoids where photosynthesis takes place. As we saw with other organelles, membrane folding increases the surface area over which biophysical processes can take place. In the case of chloroplasts, this increases absorption of light for photosynthesis.

Membrane proteins in the thylakoids form a complex with a class of molecules called *chlorophylls*. Chlorophylls absorb light and, together with the membrane proteins in the thylakoids, convert the light energy into the high-energy bonds in ATP and the chemical bond energy of carbohydrates (sugars and starches). The process requires that a magnesium ion be bound to the chlorophyll (as shown in Fig. 8-14). Several different types of chlorophylls are known. The most common and universal is chlorophyll a. Others include chlorophyll b, c1, c2, and d. The different chlorophylls absorb light of different wavelengths, but for the most part they

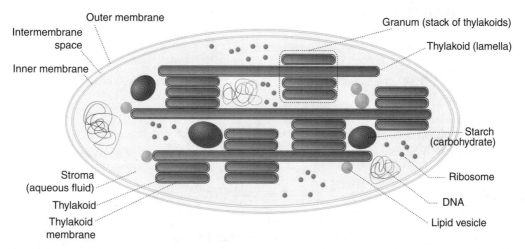

FIGURE 8-13 • Chloroplast. (*Derived from Wikimedia Commons.*)

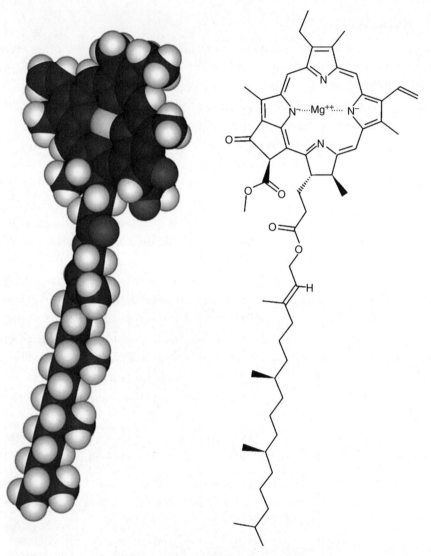

FIGURE 8-14 • Chlorophyll a: space filling model (left) and structural formula with bound magnesium ion (right).

do not absorb wavelengths in the green area of the spectrum. Since the green light is not absorbed, it is reflected; this accounts for the green color of most plants.

Cell Life Cycle

At the beginning of this chapter we noted that all cells come from preexisting cells through the process of cell division. The phases of cell growth and cell division make up what is known as the *cell cycle*. There are two main phases to

Cell cycle

Interphase: Cell growth
and DNA replication

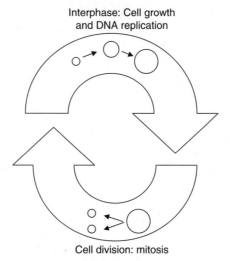

Cell division: mitosis

FIGURE 8-15 · The two parts of the cell
cycle, interphase and cell division.

the cell cycle, each of which is broken up into subphases. The two main phases
of the cell cycle are *interphase* and *cell division*. See Fig. 8-15.

There is a third phase, called *quiescent*, where the cell has essentially left
the cycle and entered into a resting phase where the cell continues to utilize
energy to do work, but has stopped dividing. The quiescent or resting phase
is abbreviated G_0.

The *interphase* of the cell cycle is where the cell puts most of its energy into
growth and DNA replication. Interphase is broken up into three subphases: G_1,
S, and G_2. During G_1 and G_2 the cell grows. Protein synthesis is very active during G_1 and G_2, and the enzymes built from this protein synthesis go on to catalyze other reactions that digest carbohydrates for energy and manufacture lipids
and other structures needed for cell growth. In between G_1 and G_2 is the
S-phase when DNA replication takes place. During the S-phase, protein synthesis is for the most part limited to those proteins that are needed for DNA
replication. For example, the enzymes involved in replication and, in eukaryotes, the structural proteins that make up histones for the packaging of chromatin. See Fig. 8-16.

After the G_2 phase, the cell enters the *cell division* phase, also called the *mitosis*
or *M-phase*. Mitosis itself is broken into several subphases. In the first stage of
mitosis, called *prophase*, the chromatin condenses into visible chromosomes.
(Remember from earlier in this chapter that sometimes DNA is spread out in a

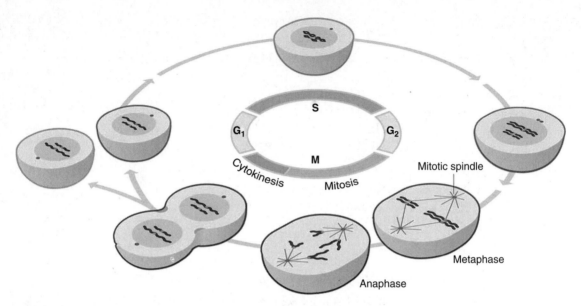

FIGURE 8-16 • Cell cycle. (*Courtesy of David O. Morgan,* The Cell Cycle: Principles of Control, *New Science Press.*)

spaghetti-like structure called *chromatin*. At other times the cell organizes individual DNA molecules into tightly packed chromosomes.) In the *metaphase* of mitosis, the chromosomes line up in pairs. A portion of the cytoskeleton known as the *mitotic spindle* begins pulling the paired chromosomes apart, so that each daughter cell will have its own set of chromosomes. After the chromosomes have separated, they begin decondensing back into chromatin. In the final phase of mitosis, called *cytokinesis*, the cell membrane pinches in toward the center of the cell and divides the cell in two. See Fig. 8-16.

Still Struggling

This chapter is largely descriptive. The main point is to learn some vocabulary and to gain some understanding of the context for the biophysical studies that we will learn about in the rest of the book. Do not feel that you have to have all of the terms in this chapter memorized. As you encounter some of these terms later in the book, use this chapter as a reference as needed, or refer to the glossary in the back of the book.

QUIZ

Refer to the text in this chapter if necessary. Answers are in the back of the book.

1. Which of the following is *not* true about cells?
   A. The cell is the basic unit of life.
   B. Cells are surrounded by a phospholipid bilayer membrane.
   C. Prokaryotes always have nucleic acids.
   D. Eukaryotes are always multicellular organisms.

2. Which of the following statements are true?
   S1. Eukaryotes are typically larger that prokaryotes.
   S2. Eukaryotes divide more rapidly than prokaryotes.
   S3. Eukaryotes compartmentalize various functions into membrane-bound organelles.
   S4. Eukaryotes adapt more quickly to drastic changes in their environment.
   A. S1 and S3
   B. S2 and S4
   C. S1 and S4
   D. S2 and S3
   E. S1, S3, and S4

3. What are receptors?
   A. Antenna-like projections from the cell that receive radio signals.
   B. Membrane proteins that receive other cells.
   C. Molecules on the surface of the cell membrane that signal specific outside molecules to come and bind to the membrane.
   D. Molecules on the surface of the cell membrane that bind specific outside molecules, thereby signaling the cell to do something.

4. What are the Golgi apparatus and the ER?
   A. The Golgi apparatus is a device used to study cellular function, and the ER is a reticulating structure found within the cell.
   B. They are both membranous organelles.
   C. The Golgi apparatus stores vesicles, and the ER stores ribosomes.
   D. The Golgi apparatus is a species of eukaryote, and the ER is a species of prokaryote.

5. What do ribosomes do for the cell?
   A. Process DNA
   B. Generate polypeptides
   C. Distinguish between the smooth and rough ER
   D. Synthesize lipids

6. **What are the two main parts of the cell cycle?**
   A. The handle bars and pedals
   B. Interphase and outerphase
   C. Growth and division
   D. Expansion and contraction
   E. Separation and division

7. **Microtubules are part of what structure?**
   A. Tubulin
   B. Mitochondria
   C. Bones
   D. Cytoskeleton
   E. Capillaries

8. **What does the nucleus of the cell contain?**
   A. Protons and neutrons
   B. The nuclear envelope and DNA lamina
   C. DNA and histones
   D. DNA and ribosomes

9. **What are chloroplasts?**
   A. Mitochondria
   B. Photosynthetic proteins
   C. Plant organelles responsible for photosynthesis
   D. Green leafy plants

10. **What do mitochondria contain?**
    A. DNA
    B. Folded membranes
    C. ATP
    D. A and B
    E. B and C
    F. A and C
    G. A, B, and C

chapter 9

Protein Biophysics

In Chap. 7 you learned some basics of protein structure and function. In this chapter we will delve a little deeper into the physics and biophysical chemistry involved. Proteins, in order to function properly, need to fold up into a specific shape. Therefore, much of protein biophysics is concerned with gaining a deep understanding for the physics of protein folding and protein conformational changes.

CHAPTER OBJECTIVES

In this chapter, you will

- Learn about protein structure and its relationship to protein function.
- Study the forces and factors that stabilize protein structures.
- Learn how membrane protein folding differs from cytoplasmic protein folding.
- Learn ways to analyze peptide bond angles.
- Learn the two most common protein secondary structures: the alpha helix and the beta sheet.

Protein Folding

When a protein is folded into its correct shape, it is said to be in its *native state*. The native state can include one or a small number of conformations that are involved in the natural functioning of the protein. When a protein is unfolded, so that it is no longer in its native state, we say that it is *denatured*.

Experimentally, we can expose proteins to *denaturing agents*, such as heat, changes in pH, or exposure to certain chemicals. Denaturing agents destabilize the protein's higher-order structure (i.e., quaternary, tertiary, and/or secondary structure) causing the polypeptide chain to unfold. Frying an egg makes the clear part of the egg become hard and white, because the protein in the egg white, albumen, becomes denatured. Frying an egg is irreversible, because the peptide chains from neighboring albumen molecules become entangled in one another. However, under carefully controlled experimental conditions, it is possible to avoid entangling of the denatured chains (e.g., by keeping the protein concentration low and using a gentle chemical denaturing agent). In such a case denaturing is reversible. We can remove the denaturing agent and allow the protein to fold back into its native state. In this way we can experimentally reproduce the folding and unfolding of a protein in order to study the physics of protein folding.

Protein folding is, for the most part, an example of *self-assembly*. Self assembly means that the forces driving the molecules to assemble in the appropriate configuration or quaternary structure are inherent in the molecules themselves, and in their interactions with solvent molecules (such as water and small ions in solution). Only molecules that are part of the final structure are involved. Solvent molecules such as water and small ions in solution *are* part of the final structure in that they remain associated with the folded protein (or other self-assembled structure) forming hydrogen bonds and other interactions that contribute to stabilizing the structure.

Not all protein folding is self-assembly. Many proteins do require assistance from other molecules, called *folding moderators* or *chaperones*, in order to achieve their native state. In some cases the folding moderators (often proteins themselves) catalyze a step in the folding process that would otherwise occur very slowly. In other cases a folding moderator may temporarily bind to the peptide chain and bring certain parts of the molecule closer together or may provide a sequestered environment that fosters the protein's native state.

Proteins are polymers of amino acids. There are 20 major amino acids commonly found in proteins. Other amino acids are also found in proteins (many of which are slight variations of the major 20), but these 20 are by far the most common. Table 9-1 lists the 20 major amino acids and some of their properties.

TABLE 9-1 The 20 major amino acids most commonly found in proteins are typically classified into the following three groups: nonpolar (hydrophobic), uncharged polar (hydrophilic), and charged (sometimes called charged polar) (hydrophilic). Among the charged amino acids there are those that are acidic (release protons and carry a negative charge) and those that are basic (absorb protons and carry a positive charge). (Courtesy of *Biochemistry Demystified*.)

Name	Abbreviation(s)	Unique Features	Structure
Nonpolar, hydrophobic side chains—found in protein interiors			
Glycine	Gly, G	• Smallest • R is a single proton • Flexible • Neuroinhibitor • Role in biosynthesis of many compounds, such as purines	$^{+}H_3N$—C—COO$^-$ with H above and H below
Alanine	Ala, A	• Methane • α–Keto homologue is pyruvate • Role in nitrogen transport from tissues to the liver	$^{+}H_3N$—C—COO$^-$ with CH_3 above and H below
Valine	Val, V	• Butyl group • Branched chain amino acid found in high concentration in muscles	CH_2 / H—C—CH_3 / $^{+}H_3N$—C—COO$^-$ / H
Leucine	Leu, L	• Branched chain amino acid found in high concentration in muscles	CH_3 / H—C—CH_3 / H—C—H / $^{+}H_3N$—C—COO$^-$ / H
Isoleucine	Ile, I	• Isomer of leucine • Branched chain amino acid found in high concentration in muscles	CH_3 / CH_2 / H_3C—C—H / $^{+}H_3N$—C—COO$^-$ / H

TABLE 9-1 The 20 major amino acids most commonly found in proteins are typically classified into the following three groups: nonpolar (hydrophobic), uncharged polar (hydrophilic), and charged (sometimes called charged polar) (hydrophilic). Among the charged amino acids there are those that are acidic (release protons and carry a negative charge) and those that are basic (absorb protons and carry a positive charge). (Courtesy of *Biochemistry Demystified*.) *(Continued)*

Name	Abbreviation(s)	Unique Features	Structure
Nonpolar, hydrophobic side chains—found in protein interiors			
Methionine	Met, M	• One of two amino acids containing sulfur • Contains a sulfur as an ester • Sulfur easily oxidized • As s–adenosyl–l–methionine (SAM), a methyl donor in many bioreactions	
Proline	Pro, P	• α–Carbon forms a ring containing primary amine • Inflexible • Forms kinks in secondary structures	
Phenylalanine	Phe, F	• Alanine plus a phenyl • Converted to tyrosine, which is, in turn, converted to l–dopa • Interferes with the production of serotonin	
Tryptophan	Trp, W	• Bulky, aromatic side chains • Indole group • Precursor for serotonin and niacin	

| TABLE 9-1 | The 20 major amino acids most commonly found in proteins are typically classified into the following three groups: nonpolar (hydrophobic), uncharged polar (hydrophilic), and charged (sometimes called charged polar) (hydrophilic). Among the charged amino acids there are those that are acidic (release protons and carry a negative charge) and those that are basic (absorb protons and carry a positive charge). (Courtesy of Biochemistry Demystified.) (Continued) |

Name	Abbreviation(s)	Unique Features	Structure
Uncharged polar side chains—metabolically active and located on the exterior of proteins			
Serine	Ser, S	• Hydroxyl group • Found at the active site of enzymes • Aids in glycoprotein formation	
Threonine	Thr, T	• Methyl and hydroxyl group • Found at the active site of enzymes • Aids in glycoprotein formation	
Asparagine	Asn, N	• Methyl group and carboxyl group with a highly polar uncharged amine that readily forms hydrogen bonds • Found at the ends of alpha helices and beta sheets • Aids in formation of glycoproteins • Input to urea cycle • α–Keto homologue is oxaloacetate	
Glutamine	Gln, Q	• Two methyl groups and carboxyl group with a highly polar uncharged amine • Forms isopeptide linkages • Central role as nitrogen donor in synthesis of nonessential amino acids • Provides nitrogen transport to the liver	

TABLE 9-1 The 20 major amino acids most commonly found in proteins are typically classified into the following three groups: nonpolar (hydrophobic), uncharged polar (hydrophilic), and charged (sometimes called charged polar) (hydrophilic). Among the charged amino acids there are those that are acidic (release protons and carry a negative charge) and those that are basic (absorb protons and carry a positive charge). (Courtesy of *Biochemistry Demystified*.) *(Continued)*

Name	Abbreviation(s)	Unique Features	Structure
Uncharged polar side chains—metabolically active and located on the exterior of proteins			
Tyrosine	Tyr, Y	• Similar to phenylalanine but with polar hydroxyl group on phenyl ring • Important metabolically because ionization altered by micro pH changes	
Charged polar side chains—reactive			
Cysteine	Cys, C	• Sulfhydryl (thiol) group • Forms disulfide bridges • Found at the active site of enzymes • Binds iron	
Lysine	Lys, K	• Long aliphatic chain terminating in an amine • Nucleophilic • Forms ionic bonds	
Arginine	Arg, R	• Long aliphatic chain containing an amine and terminating in two amines • Nucleophilic • Forms ionic bonds • Generated in the urea cycle	

TABLE 9-1 The 20 major amino acids most commonly found in proteins are typically classified into the following three groups: nonpolar (hydrophobic), uncharged polar (hydrophilic), and charged (sometimes called charged polar) (hydrophilic). Among the charged amino acids there are those that are acidic (release protons and carry a negative charge) and those that are basic (absorb protons and carry a positive charge). (Courtesy of *Biochemistry Demystified*.) *(Continued)*

Name	Abbreviation(s)	Unique Features	Structure
Charged polar side chains—reactive			
Histidine	His, H	• Methyl • Imidazole group • Ionic bonds found at the active site of enzymes • Crucial in the structure of hemoglobin	
Aspartate or aspartic acid	Asp, D	• Donates an amine to become oxaloacetate • Active in proteolytic enzymes	
Glutamate or glutamic acid	Glu, E	• Central role as nitrogen donor in synthesis of nonessential amino acids • Provides nitrogen transport to the liver	

You can see the side chains vary significantly among the 20 major amino acids. Some, like tryptophan, are large, creating steric forces that limit the possible conformations available to the peptide chain. Others, like glycine and alanine, are small, allowing the possibility of tightly packed conformations.

Some amino acid side chains are hydrophilic, either uncharged polar or charged (ionized) at neutral pH. Others are nonpolar and hydrophobic. Three of the 20 amino acids are aromatic. Phenylalanine and tryptophan are aromatic and hydrophobic, whereas tyrosine is aromatic but polar owing to a hydroxyl group on the aromatic ring.

Factors Influencing Protein Structure

In this section we list and discuss the factors influencing protein structure. These factors include the forces affecting the conformation of biomolecules (Chap. 6) as they apply specifically to amino acids and peptide chains. We begin with a discussion of charge interactions. Charges occur in proteins due to the ionization of the amine and carboxyl groups of amino acids at the ends of the polypeptide chain, or due to the ionization of various amino acid side chains. Charges cause the protein to fold in a way that maximizes the distance between like charges, while bringing opposite charges as close together as possible, all of which minimizes the free energy and stabilizes the protein structure.

Ionization of Amine and Carboxyl Groups

The functional groups that define an amino acid are the amine and carboxyl groups on the α-carbon. Both of the groups are easily ionized at physiological pH, the amine group by the addition of a proton, and the carboxyl group by the loss of a proton. See Fig. 9-1. The result is a *zwitterion*, an ionized molecule that has both a positive and a negative charge on it, but a net charge of zero.

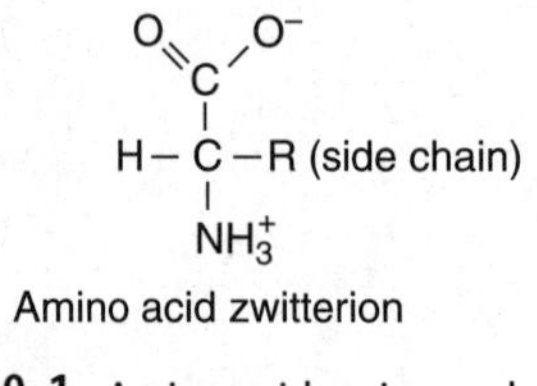

Amino acid zwitterion

FIGURE 9-1 • Amino acid amine and carboxyl groups are easily ionized.

In a polypeptide chain, the amine and carboxyl groups participate in the covalent peptide bonds, eliminating the possibility of ionizing these groups except at the ends of the chain. Nonetheless that still means each polypeptide chain in a protein will have a positive charge at one end and a negative charge at the other; this can influence the folded structure of the protein. Most proteins consist of a single polypeptide chain; however, many proteins contain more than one polypeptide chain bound together in a quaternary structure.

Ionization of Amino Acid Side Chains

Five of the 20 major amino acids have side chains that carry a charge (ionize) at physiological pH. Notably aspartate and glutamate each carry a negative charge, while histidine, lysine, and arginine carry a positive charge (see Table 9-1).

Still Struggling

What's the difference between polar versus charged? When we use the word charged, we mean that the molecule is ionized. Ions contain a unit charge (the amount of charge on an electron or proton) or they contain a multiple of a unit charge. In contrast, the word polar is used in various ways. Strictly speaking, polar means the presence of *both* positive and negative charges on the same molecule, but these can be either unit charges or partial charges. In the case of an ionized molecule, polar would imply that the molecule is a zwitterion. But ionized amino acid side chains typically contain only a positive or negative charge. Still, scientists sometimes refer to such single charge ions as polar, because ions prefer (energetically) to be in a polar environment—that is, an environment containing polar molecules such as water. Another use of the word polar is this: we say that a molecule or bond is polar when the electrons are distributed unevenly, so there is some charge, but it is a partial charge. According to this use of the word polar, *charged side chains contain more charge than polar side chains*. Recall from Chap. 6 that the attractive or repulsive force between two charges (and the energy of their interaction) is proportional to product of the amount of each charge. Therefore the Gibbs energy change associated with the interaction of charged (ionized) side chains is stronger than that associated with polar side chains.

These charged amino acid side chains can influence protein structure in a number of ways. They can form salt bridges. A *salt bridge* is a single ionic bond between a positive and negative ion. In addition to salt bridges that hold parts of the molecule together, like charges will repel and tend to push certain parts of the polypeptide chain away from each other. Positive charges can participate in a cation-pi attraction with an aromatic side chain (see Chap. 6). Similarly a negative charge will result in repulsion from the side of an aromatic group, causing negatively charged side chains to be positioned away from aromatic side chains.

Polar Amino Acid Side Chains

At least a half-dozen of the 20 major amino acids contain uncharged polar, hydrophilic side chains. These side chains will tend to face toward the outside of the protein in an aqueous environment. They will also tend to be situated near other polar side chains and charged side chains as well. Polar side chains often participate in hydrogen bonding, both with water and with other polar side chains. In a membrane protein, the polar amino acid side groups will tend to fold toward the polar outer edges of the lipid bilayer, and may even protrude out of the bilayer.

Hydrogen Bonds in Proteins

Studies have shown that, in an average protein, nearly 90% of the functional groups that can form hydrogen bonds do form hydrogen bonds, either with water or with other functional groups. Despite the abundance of hydrogen bonds, the difference in energy between two amino acid side chains hydrogen bonding with each other versus each side chain forming a hydrogen bond with water, is small. This reasoning and evidence from chemical studies that disrupt hydrogen bonds (without disrupting other forms of van der Waals forces), indicate that although hydrogen bonds play a role in stabilizing the structure of proteins, they are not the greatest contributing factor. Equation (9-1) shows the equilibrium between two amino acid functional groups, call them X and Y, hydrogen bonding with each other versus hydrogen bonding with water. The hydrogen bonds are shown as dotted lines. On the left side of the equation X and Y hydrogen bond with each other, and water hydrogen bonds with water. On the right side of the equation X hydrogen bonds with water and Y hydrogen bonds with water. The energy difference between the two sides of the equation is small (although it may vary depending on the exact environment of the groups). Both types of hydrogen bonds form off and on dynamically. Thus the hydrogen bonds between different parts of the polypeptide chain only contribute slightly to the stability of the protein structure.

$$XH \text{ --- } Y + HOH \text{ --- } OH_2 \leftrightarrow XH \text{ --- } OH_2 + Y \text{ --- } HOH \qquad (9\text{-}1)$$

Nonpolar Amino Acid Side Chains

Nonpolar amino acid side chains are a source for the hydrophobic effect. As explained in Chap. 6, hydrophobic functional groups create a free-energy penalty by forcing water molecules adjacent to the hydrophobic molecule into a more ordered state and lowering their entropy. This means there is a free-energy benefit

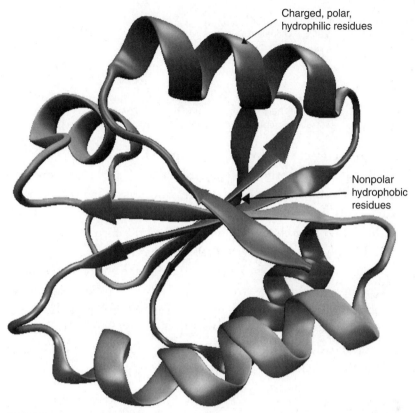

Charged, polar,
hydrophilic residues

Nonpolar
hydrophobic
residues

FIGURE 9-2 · Schematic diagram of a folded protein polypeptide chain, with hydrophobic portions on the inside and polar and charged groups on the outside. (*Courtesy of Wikimedia Commons.*)

to moving these functional groups away from water, into the center of the protein. Studies show that while all of the factors mentioned in this section contribute energetically to stabilizing protein structure, the hydrophobic effect contributes the most. Proteins are easily denatured in the presence of hydrophobic solvents that weaken the hydrophobic effect. Cooler temperatures also weaken the hydrophobic effect, by ordering all of the water molecules so that the cost of ordering the water molecules adjacent to a hydrophobic functional group becomes negligible. Figure 9-2 illustrates a folded protein with the hydrophobic functional groups on the inside and polar and charged functional groups on the outside.

Disulfide Bonds

Two of the major amino acids contain sulfur. One in particular, cysteine, contains a sulfur atom at the end of its functional group with only a small hydrogen atom attached. This makes that sulfur atom especially reactive, and there is a

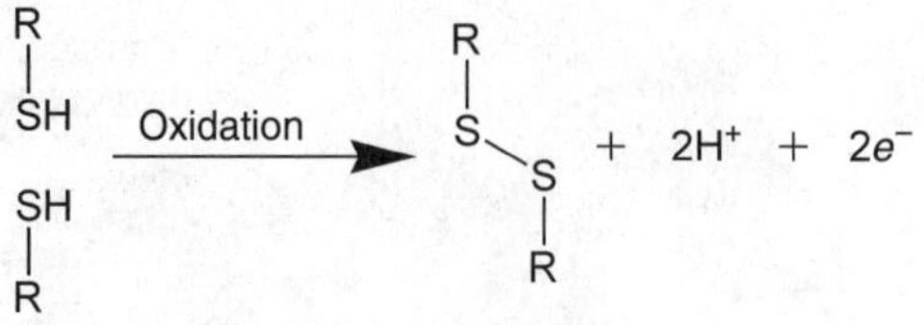

FIGURE 9-3 · Disulfide bonds between cysteine residues help stabilize and add rigidity to protein structure. (*Courtesy of Wikimedia Commons.*)

favorable free-energy change for this sulfur atom to form a covalent bond with another sulfur atom on another cysteine residue. Such a covalent bond between two sulfur atoms, attaching two cysteine residues together, is called a *disulfide bond*. Many proteins contain cysteine residues some distance away from each other in the amino acid chain. Figure 9-3 shows the formation of a disulfide bond. Disulfide bonds contribute favorably to the free energy of protein folding. In addition, once formed they add some rigidity to a protein secondary and tertiary structure, enabling the structure to withstand a significant amount of bumping about and stress while still maintaining its native state. The formation of disulfide bonds is a type of *cross-linking* (Chap. 6) a covalent connecting of parts of a molecule that are distant in primary structure but come together (and cross-link) in the secondary or tertiary structure.

Peptide Bond Dipoles

The amide group and carbonyl group on the polypeptide backbone are each polar. The uneven charge distribution of the electron cloud means that each group is effectively a dipole. The amide group has a charge of approximately 0.2 and the carbonyl approximately 0.4. Together, in a peptide bond, these two dipoles act as a single dipole with D (the dipole moment) equal to about 0.7. See Fig. 9-4.

Since a protein (a polypeptide) is a whole chain of such dipoles strung together, you can see how this could have a significant effect on the conformation of the protein. Recall from Chap. 6 that the energy of dipole-dipole interactions varies

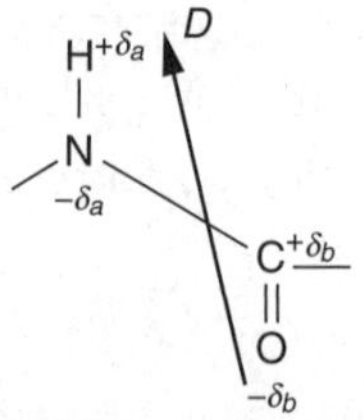

FIGURE 9-4 · Peptide bond as a dipole. The partial charge $\delta_a = 0.2$ and $\delta_b = 0.4$; the dipole moment $D = 0.7$ for the entire peptide bond including the amide and carbonyl groups.

as $1/r^3$, where r is the distance between the dipoles. The angle of orientation between any two dipoles also has a significant effect on the free energy. Dipoles lined up in parallel will repel each other, while those lined up antiparallel will attract one another. Furthermore dipoles lined up end to end, with the positive end of one dipole facing the negative end of another dipole, will effectively see each other as point charges. The energy associated with point charges varies as $1/r$ and is thus much stronger and can be felt over a larger distance of a typical dipole-dipole interaction. Thus protein conformations in which the peptide bond dipoles line up end to end (or close to end to end) may be favored over other conformations. This is highly significant for certain protein structures, for example the alpha helix structure which we discuss below.

Membrane Proteins

Some proteins function within membranes. Within the membrane, proteins play various roles. Structural proteins anchor the cytoskeleton to the membrane. Some membrane proteins act as receptors allowing the cell to bind to specific objects (or specific objects to bind to the cell). Enzymes often catalyze reactions within or on the surface of membranes. And transport proteins aid the movement of ions through the cell membrane, often with the goal of maintaining an electric potential difference on opposite sides of the membrane.

Membranes are lipid bilayers in which the majority of the bilayer (the inside) is very hydrophobic, with a layer of charged and polar hydrophilic groups making up both surfaces of the membrane. Although the factors that influence protein folding are the same, in practice the end result of folding a membrane protein is quite different from that for proteins found in the aqueous environment of the cytoplasm. Membrane proteins often fold in such a way as to span the membrane, with hydrophobic residues in the membrane center and hydrophilic residues exposed to the outer surface (see Fig. 9-5). In some membrane proteins, particularly those which must transport ions and other hydrophilic molecules from one side of the membrane to the other, the portion of the protein that spans the membrane is folded in a way exactly opposite as that of nonmembrane proteins. That is, we find hydrophobic residues on the outside of the protein, with hydrophilic residues on the inside. The hydrophilic residues on the inside of such a membrane protein create a protective tunnel, allowing ions and other small hydrophilic molecules to pass through the membrane. Otherwise the energy required to move an ion or hydrophilic molecule into and through the cell membrane would be prohibitive.

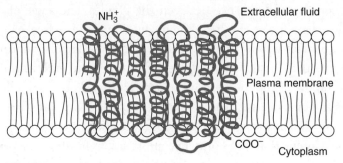

FIGURE 9-5 · Membrane proteins fold in a way as to place hydrophobic residues in the center of the membrane and hydrophilic residues at the membrane surface. Sometimes, for example in the case of proteins called *ion channels,* some hydrophilic residues fold up to the *inside* of the portion of the protein that spans the bilayer, creating a tunnel through which ions and other small hydrophilic molecules can pass.

Analysis of Polypeptide Backbone Bond Angles

The polypeptide backbone requires a good deal of flexibility in order for a polypeptide to fold up into a globular protein. Although side chain flexibility has some effect on protein conformation, the backbone ties the entire chain together and therefore determines or limits the possible shapes that a polypeptide can take. Figure 9-6 shows a tripeptide (a three-residue polypeptide) in order to illustrate the features of the polypeptide backbone. The backbone consists of a repeating chain of three covalent bonds. These bonds are

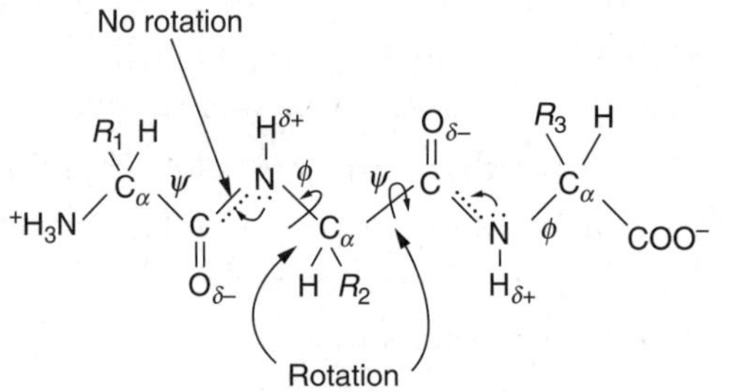

FIGURE 9-6 · A tripeptide is used to illustrate the features of the polypeptide backbone. The polypeptide backbone consists of three repeating covalent bonds, two of which (the N-Cα and Cα-C bonds) allow rotation. The third bond (C-N) does not allow rotation. The dotted lines indicate the partial double bond character of the C-N bond. The two dots next to each nitrogen indicate the otherwise non-bonded nitrogen electrons that spend a portion of their time within the C-N bond.

1. Between the amide nitrogen and the α-carbon (angle of rotation = ϕ)

2. Between the α-carbon and the carbonyl carbon (angle of rotation = ψ)

3. Between the carbonyl carbon and the amide nitrogen (no rotation)

The third bond, the one between the carbonyl carbon and the amide nitrogen, cannot rotate. This is because the pair of nonbonded electrons on the nitrogen spends a portion of its time within that bond (so they are not entirely nonbonding once they are part of a peptide chain). This gives the bond somewhat of a double-bond character. This in turn prevents free rotation around the bond. It also means that the amide nitrogen, the carbonyl carbon, and the carbonyl oxygen must all lie within a plane.

The other two bonds, however, allow free rotation, and we use the symbols ϕ and ψ respectively to represent the angle of rotation between these bonds and a plane defined by the carbonyl group and the amide nitrogen.

The N-Cα and Cα-C bonds themselves allow free rotation; however, other forces come into play to limit the possible values of ϕ and ψ. Obviously all of the forces mentioned above can come into play (electric charge, hydrophobicity, etc.), but the two biggest contributors to the limits on ϕ and ψ are (1) steric hindrances and (2) dipole-dipole interactions from adjacent peptide bonds. By taking these two forces into account, it is possible to calculate the potential energy (free energy) associated with every possible pair of ϕ and ψ values. Steric hindrances alone will disallow many values of ϕ and ψ depending on which amino acid side chains are considered. Taking dipole interactions into consideration refines the calculations. If we then plot a set of ϕ and ψ for which the energy is within a particular range, we get a contour diagram that might look similar to Fig. 9-7. A ϕ, ψ contour diagram is also called a *ramachandran diagram* for the biophysicist G. N. Ramachandran who invented it.

Common Protein Secondary Structures

There are two very common secondary structural motifs that occur throughout protein structures. One is called the *alpha helix* and the other is called the *beta sheet*. Other structural motifs do occur, but these two are the most common. Any given globular protein (tertiary structure) can contain various sections of alpha helix and beta sheet, as well as other structures, along the polypeptide chain. This can be seen in Fig. 9-2.

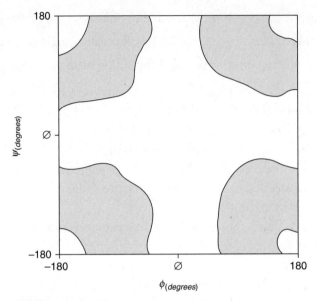

FIGURE 9-7 · An example of what a ϕ, ψ contour diagram might look like, depending on which amino acid side chains are considered. The shaded region shows the values for which ϕ and ψ result in a negative free energy. In other words these values of ϕ and ψ result in an energetically stable conformation. The unshaded regions represent ϕ and ψ values that are prohibited due to either steric hindrances or an unfavorable (repulsive) dipole interaction.

Still Struggling

A ϕ, ψ contour diagram is just a convenient way to visualize which bond angles are favorable and which are not. The shaded area represents the favorable values of ϕ and ψ. The larger the shaded area, the more favorable possibilities exist for the bond angles ϕ and ψ. This means the larger the shaded area, the more flexible is the polypeptide chain. Knowing the structure and shape of each amino acid, we can calculate a ϕ, ψ contour diagram for any pair of adjacent amino acids. Some will have a lot of shaded area. Others will have less. In this way we can get some idea as to how the amino acid sequence affects the flexibility of a protein.

Alpha Helix

The *alpha helix* is a secondary structure in which the polypeptide chain forms a right-handed helical shape. The backbone of the chain is on the inside of the helix with the various amino acid side chains on the outside. (This is in contrast to DNA secondary structure in which the backbone is on the outside of the helix and the nucleotide bases are on the inside.)

One of the characteristic features of the alpha helix is that it maximizes the number of hydrogen bonds that can form among the amide and carbonyl groups of the polypeptide backbone. The hydrogen atom from each of the backbone amides (N-H) forms a hydrogen bond with the peptide carbonyl oxygen (C = O) that is four residues further along on the chain. The result is 3.6 amino acid residues, or peptide bonds, per turn of the helix. The pitch of the alpha helix is 54 nm; that is, each turn of the helix is about 54 nm long (or 15 nm per amino acid residue). The backbone bond angles are $\phi = -57°$ and $\psi = -47°$. Figure 9-8 shows a schematic diagram of an alpha helix. Some of the hydrogen bonds are shown as dotted lines.

The alpha helix is also a conformation that maximizes the energy benefit from the peptide bond dipoles (Fig. 9-4). Adjacent peptide bond dipoles are somewhat side by side, so their dipole interaction (which varies as $1/r^3$) is repulsive. But peptide bond dipoles further along the helix are lined up nearly end to end (see Fig. 9-9). As explained above, end-to-end dipole-dipole interactions are approximately equivalent to charge-charge interactions. This makes them significantly stronger over longer distances (compared with the adjacent

FIGURE 9-8 · Diagram of a polypeptide alpha helix.

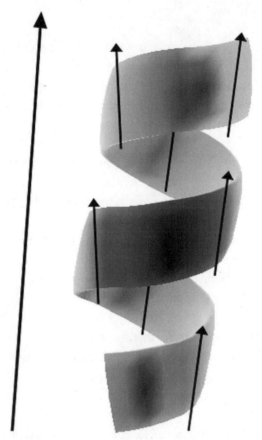

FIGURE 9-9 • Peptide bond dipoles contribute to the stability of the alpha helix.

peptide dipoles) since their interaction energy varies as $1/r$. The result is that overall in the alpha helix dipole-dipole attractions outweigh the dipole-dipole repulsions. This means that overall the peptide bond dipoles contribute to stabilizing the helix. They also cause the entire helix to behave as one large dipole, as shown schematically in Fig. 9-9. In this way alpha-helical portions of globular proteins can interact with each other as dipoles, thus affecting tertiary conformation. The alpha-helical parts also interact as dipoles with other molecules and proteins while carrying out protein functions (binding, catalysis, etc.). Helical segments within proteins range anywhere from 3 to 40 or more residues in length, but about 10 residues is average. Proteins, however, can have many alpha-helical segments separated by bends or other secondary structures.

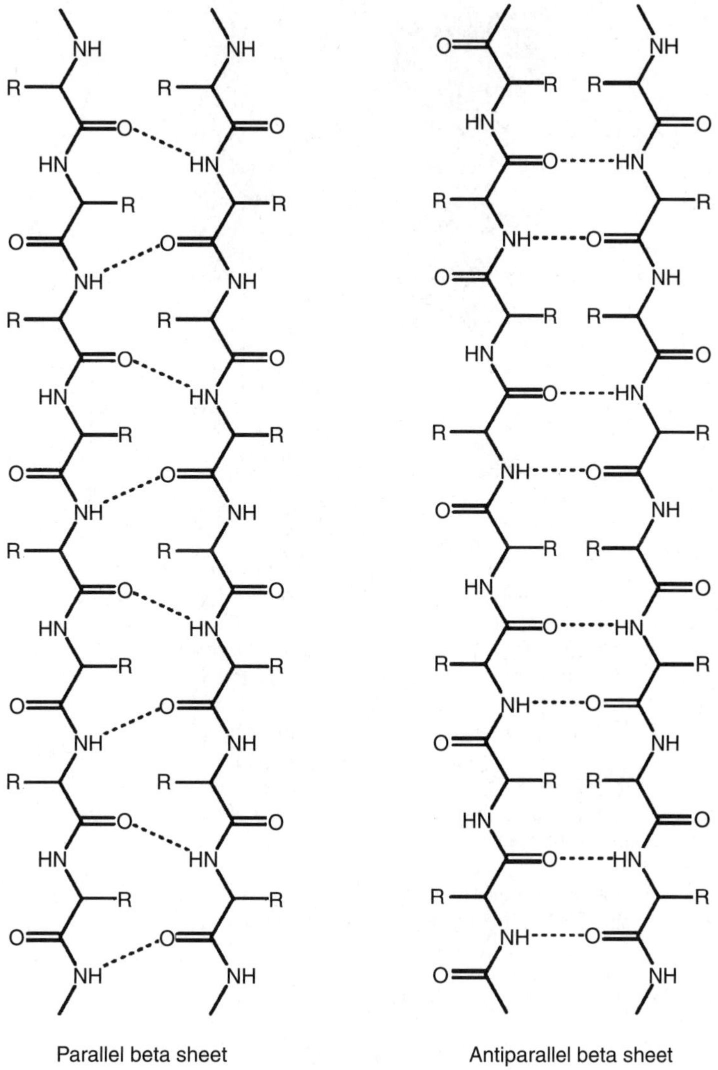

Parallel beta sheet   Antiparallel beta sheet

FIGURE 9-10 • Beta sheet polypeptide secondary structure.

Beta Sheet

The *beta sheet* (denoted as *β sheet* and also called *beta pleated sheet*) is the second most common secondary structure found in proteins. A beta sheet consists of two polypeptide strands with their backbones hydrogen bonded to one another. The two strands are called *beta strands* and are actually one strand of a single polymer that has folded back on itself. A beta strand then

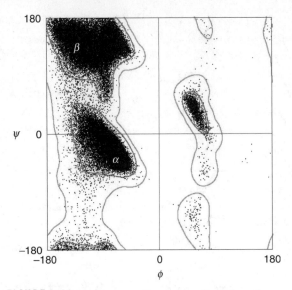

FIGURE 9-11 · ϕ, ψ contour diagram showing where the alpha helix (α) and beta sheet (β) fall on the ϕ,ψ plot. Each black dot represents a residue in a protein. Each contour line represents a level of potential energy, with lower energies represented by the inner lines and higher energies by the outside lines. Notice that the alpha helix and beta sheet fall in areas of very low potential energy, indicating that they are both very stable secondary structures. Notice also that the majority of residues (black dots) fall close to the helix and beta sheet, indicating that most of the residues of the protein are in one of these two configurations. The small, darkly shaded area on the right side represents favorable bond angles for residues in configurations other than alpha helix and beta sheet. (*Courtesy of Wikimedia Commons.*)

is a stretch of amino acids within a polypeptide. Each beta strand is typically five to ten residues long. The key here is that the hydrogen bonds between the amide hydrogen atoms and the carbonyl oxygen atoms are from one strand to another (as opposed to within the same strand, four residues away, as was the case with the alpha helix). The two strands next to one another create a sheetlike effect. In the alpha helix the carbon bond angles are tighter and twisted. But in the β sheet configuration, these bond angles are more stretched out, creating a back-and-forth pattern that gives the β sheet a pleated appearance.

Beta sheets can be parallel or antiparallel. As mentioned, the polypeptide backbone consists of a repeating chain of three covalent bonds, or three atoms: (1) the amide nitrogen, (2) the α-carbon, and (3) the carbonyl-carbon. If we denote the α-carbon as C^α, and the carbonyl-carbon as C', then a parallel beta sheet means that as we move along the sheet, both strands have the backbone atoms in the same order. For example: N, C^α, C', N, C^α, C'. Antiparallel means that the atoms on one strand go in the order N, C^α, C'; the atoms on the other strand go in the order N, C', C^α. Figure 9-10 shows hydrogen bonding in parallel and antiparallel beta sheets.

QUIZ

Refer to the text in this chapter if necessary. Answers are in the back of the book.

1. The 20 major amino acids are typically classified into what three groups?
    A. Nonpolar, uncharged polar, and charged
    B. Hydrophobic, hydrophilic, and hydroponic
    C. Acidic, basic, and aromatic
    D. Side chains, linear, and circular

2. Which two polar bonds contribute to making a peptide bond polar?
    A. The α-carbon-H and the carbonyl C = O
    B. The amide N-H and the α-carbon-H
    C. The amide N-H and the carbonyl C = O
    D. The α-carbon-H and the disulfide S-S

3. Which has the largest influence on stabilizing protein structure?
    A. Zwitterions
    B. Hydrophilic dipoles
    C. Hydrophobic forces
    D. Hydrogen bonds

4. Which of the following is *not* true?
    A. Membrane protein structure is inside-out relative to cytoplasmic proteins.
    B. In an aqueous environment, nonpolar residues fold into the center of a protein.
    C. Aromatic side chains are most likely to hydrogen-bond with water.
    D. A lower potential energy means the conformation is more stable.

5. What are disulfide bonds?
    A. Hydrogen bonds between sulfur atoms
    B. Covalent sulfur-sulfur bonds between two methionine residues
    C. Covalent sulfur-sulfur bonds between two cysteine residues
    D. Ionic sulfur-sulfur bonds between two cysteine residues

6. A ϕ, ψ contour diagram is based on
    A. plotting energy levels for various values of bond angles ϕ and ψ.
    B. energy due to steric hindrances and dipole interactions.
    C. both of the above
    D. none of the above

7. A beta sheet is pleated because
    A. the bond angles in the polypeptide backbone alternate back and forth.
    B. each beta strand folds back on the polypeptide chain.
    C. the alpha helix is more stable.
    D. hydrophobic residues alternate with hydrophilic residues.

8. **In an alpha helix**
   A. the backbone is on the outside.
   B. the side chains have dipoles.
   C. the dipoles are every fourth residue.
   D. the hydrogen bonds are every fourth residue.

9. **What are antiparallel strands?**
   A. One strand is oriented at 90° to the other.
   B. One strand goes up and the other strand goes down.
   C. One strand is an alpha helix, and the other is a beta helix.
   D. One strand is a double helix and the other is a beta sheet.

10. **Cooperative protein folding means that**
   A. several proteins cooperate to fold each other.
   B. the protein is either all alpha helix or all beta sheet.
   C. the alpha helix and the beta sheet stabilize one another.
   D. the protein folds into its native state all at once.

chapter 10

Nucleic Acid Biophysics

In Chap. 7 you learned some basics of nucleic acids structure and function. In this chapter we dig deeper into nucleic acid biophysics. Of the two main types of nucleic acids in nature, DNA (deoxyribonucleic acid) and RNA (ribonucleic acid), more attention is paid to DNA because of its central importance to genetics. Still the role of RNA is critical, and we have a lot to learn about the biophysics of nucleic acids from both.

CHAPTER OBJECTIVES

In this chapter, you will

- Learn about DNA and RNA structures.
- Come to understand the relationship between nucleic acid structure and function.
- Study the forces and factors that stabilize nucleic acid structures.
- Learn how energy can affect genetic functioning.

Introduction

The first thing to understand from a biophysics point of view is that DNA is more than a storehouse of genetic information, more than just a book of instructions to be read for building proteins. True, sequences of nucleotides in DNA are transcribed and translated into sequences of amino acids in proteins. But the sequences of nucleotides in DNA and RNA do more than just specifing the order of amino acids in proteins. For one thing, the nucleotide sequence of a nucleic acid has a major impact on secondary and tertiary structures, which in turn play an essential role in genetic functioning. Nucleotide sequence also influences the Gibbs energy of helix formation, affecting the winding and unwinding of the DNA double helix, which is necessary for transcription and replication. And sections of nucleotide sequence may define binding sites so that certain proteins will bind to the nucleic acid only where that specific nucleotide sequence exists.

DNA Secondary Structure

Let's start by taking a look at a space-filling model for the DNA double helix. Figure 10-1 shows a space-filling model for the B form of DNA double helix. B-DNA is the classic double helix proposed by Watson and Crick. The B-DNA double helix itself is characterized by the following properties. The width of the helix is 23.7 Å (1 Å = 1 angstrom = 10^{-10} m). The *pitch* of the helix (the length of one turn of the helix) is 33.2 Å. Each turn of the B-DNA double helix contains about ten base pairs. Combined with the pitch of the helix, this means that there is a distance of about 3.3 Å from each base pair to the next.

There are a few other points to note about the DNA double helix. First, in contrast to the alpha helix of proteins, the *backbone* of the DNA double helix is on the outside of the helix. The backbone here consists of a repeating sequence of alternating sugar and phosphate residues. In nucleic acid double-helical conformations, the base pairs are on the inside of the helix, with the sugar-phosphate backbones on the outside. Also, strictly speaking there are two backbones in a double helix, one from each nucleotide strand.

Another thing to note is that there are two helical grooves that wind around the helix, between where the backbones stick out. One grove is about 50% wider than the other, and thus they are called the *major groove* and the *minor groove*. The major groove is particularly important because it somewhat exposes the interior of the helix; this benefits proteins and other molecules that need to bind to a specific sequence of nucleotides. The major groove allows these molecules to "see" the nucleotide sequence on the inside of the double helix.

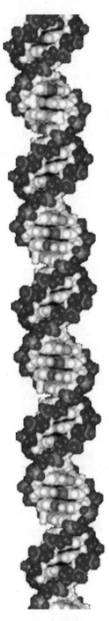

B-DNA

FIGURE 10-1 · DNA double helix in the B-form conformation.

Current evidence indicates that B-DNA is the most common DNA double-helix conformation in nature, but other conformations of double helix exist too. The most studied double-helix conformations (other than B-DNA) are A-DNA and Z-DNA, which differ from B-DNA in their helical pitch, tilt of their base pairs, and so on. Z-DNA is also different because it is a *left-handed* helix, which means that as you move along the length of the helix, the helix turns in a *counterclockwise* direction (as opposed to a *right-handed* helix, in which the helical turns are in a *clockwise* direction).

Table 10-1 lists various parameters which distinguish the different double helix conformations for A-, B-, and Z-DNA. You can also see, in Fig. 10-2, space-filling models for A- and Z-DNA, which can be compared with that for B-DNA in Fig. 10-1. Notice that the relative sizes of the major and minor groove are different. The Z-DNA conformation, in addition to being left-handed, is a much tighter conformation, resulting in a narrower helix and more base pairs per turn. The data listed in Table 10-1 are for crystalline (solid) DNA. Direct conformational measurements are best obtained using X-ray diffraction which requires molecules to be crystallized. However, experiments with DNA in solution indicate that the *helical repeat* for B-DNA in solution is about 10.4 base pairs per turn of the helix. This would mean that the base pairs are slightly closer together in solution than they are in the DNA crystal.

TABLE 10-1 Some geometric parameters describing the A-, B-, and Z- conformations of DNA double helix.

Geometric Parameter	B-DNA	A-DNA	Z-DNA
Pitch of helix (length of one turn)	33.2 Å	24.6 Å	45.6 Å
Width of helix	23.7 Å	25.5 Å	18.4 Å
Average base pairs per turn of helix	10.0	10.7	12

The A-form double helix is also the conformation observed for RNA-DNA hybrids, that is, where one strand is DNA and the other strand is RNA. Such hybrid, double-stranded nucleic acids exist temporarily during transcription. This has led to some speculation that in the cell the DNA double helix may assume the A-conformation just prior to transcription. Other, less well-studied, double-helix conformations have been shown to exist under appropriate conditions, for example, C-DNA. In double-helical DNA, the C form is favored only at very low ionic strength. It is therefore believed that C-form DNA is unlikely to occur in living cells. Many other double-helix conformations have been demonstrated to exist under varying conditions. In all, about 20 different conformations of DNA double helix have been described and named, using up most of the letters of the alphabet. However, in contrast to A-, B-, and Z-DNA, most of

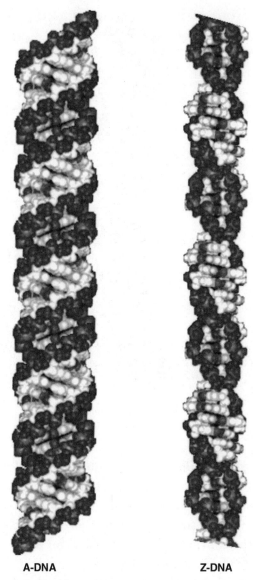

A-DNA   **Z-DNA**

FIGURE 10-2 • DNA double helix in the A-form and
Z-form conformation.

these additional conformations are only observed in synthetic DNA molecules,
under laboratory conditions, and are currently not believed to occur in nature.

Local DNA Secondary Structures

So far we have focused on DNA secondary structure in terms of various conformations of double helix. But in DNA we find also other secondary structures.

FIGURE 10-3 • Bent double helix, where B-DNA meets A-DNA.

These structures are often found in combination with the double helix, and are typically localized to a specific location along the double helix. The simplest of these is a sharp bend or kink in the double helix. Such a bend or kink might occur, for example, where a stretch of double helix in one conformation meets a stretch of double helix in another conformation, for example, where B-DNA meets Z-DNA or A-DNA; See Fig. 10-3.

In some cases the nucleotide sequence itself might create a propensity toward a curved or bent double helix. How so? Much of the energy required to stabilize the double helix comes from base-stacking interactions. But the stacking interactions between various nucleotide bases are not all equal. For example, the free-energy change of stacking an adenine on top of a thymine is −0.78 kcal/mol. Whereas the free-energy change from stacking adenine on top of guanine is about −1.19 kcal/mol. Both are negative and so both are stabilizing interactions, but clearly the A on top of G interaction is more stabilizing than A on top of T. It takes a lot more energy to disrupt an AG stack than and AT stack.

In solution, such as in the cell, molecules are constantly moving about and bumping into each other. Such bumps are more likely to temporarily disrupt or spread an AT stack than an AG stack. Therefore, over time, the average distance between adjacent adenine and guanine residues will be less than the distance between adjacent adenine and thymine residues. In other words, the more negative Gibbs energy change for an AG stack represents a stronger attraction due to the aromatic stacking interaction, and therefore the bases are on average closer together. If the nucleotide sequence is such that, as we wind around the helix, most of the stronger base stacking interactions occur on one side of the double helix, and most of the weaker base-stacking interactions occur on the other side of the helix, then the double helix will tend to bend toward the side with stronger base-stacking interactions (where the bases are on average closer together) and away from the side with the weaker stacking interactions (where the bases are on average further apart). See Fig. 10-4.

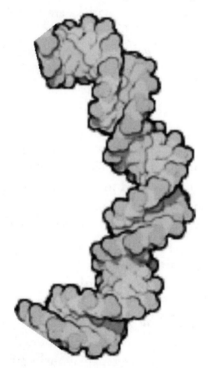

FIGURE 10-4 • Curved double helix. If the nucleotide sequence is such that, at least for some portion of the double helix, there are stronger base-stacking interactions on one side of the helix, then the helix will bend toward the bases where the stronger base-stacking interactions pull adjacent bases closer together.

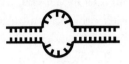

(a) Single-stranded loop   (b) Bubble

FIGURE 10-5 · Some localized DNA secondary structures.
(a) A single-stranded loop results if a portion of the
complementary nucleotide sequence is missing from the
opposite strand. (b) A bubble occurs when a portion of
the double helix unwinds.

Other localized double-helix secondary structures include single-stranded loops, bubbles, hairpins, stem-and-loop formations, and cruciform structures. A *single-stranded loop* can form within the double helix where a portion of the nucleotide sequence on one strand is not complementary to the other strand, but the nucleotides surrounding that portion are complementary to the other strand. This is illustrated in Fig. 10-5a.

As a secondary structure, a *bubble* is simply a portion of double helix that is unwound (see Fig. 10-5b). Bubbles can form in two ways. A portion of the double helix can be permanently unwound if that portion of the double helix contains bases that are not complementary between the two strands. This is unusual and not often found in nature. More often, however, although the two strands of DNA are entirely complementary, a portion of the double helix can unwind temporarily forming a bubble. This temporary secondary structure can be recognized by proteins that bind to DNA and thus serve a biological purpose. Unwound portions of the double helix are probably the most common biologically significant secondary structures found in DNA.

Hairpin, stem-and-loop, and cruciform structures can occur when a nucleotide sequence contains a *palindrome*. In molecular biology, a palindrome is defined a bit differently than it is for words or sentences. A word or sentence is considered a palindrome if reading it forward or backward gives the same result, for example the word *racecar* is a palindrome, as is the word *rotator*. However, in molecular biology we take into account the *complement* of the nucleotide sequence, that is, the sequence of nucleotides that base-pairs with a given nucleotide sequence. A *nucleic acid palindrome* is a nucleotide sequence that *is its own complement when read backward*. For example, consider the sequence

AATGCGTGGTACCACGCATT

The complement of this sequence is the sequence of bases that pair with each of the bases listed. T pairs with A, and vice versa, and C pairs with G, and vice versa. So the complement is

TTACGCACCATGGTGCGTAA

But notice that this is just the same sequence read backward!

Nucleic acid palindromes can be contiguous, as in this example, or can contain intervening sequences that are not part of the palindrome. For example,

ACGCACCATGCTGTTTGGTGCGT

has the palindrome portion of the sequence is shown shaded. The intervening sequence, TGCTGTT, is not part of the palindrome.

Palindrome sequences do occur quite often in nature, at least much more often than one might expect if an organism's nucleotide sequence were entirely random. The types of secondary structures that palindromes can form depend on whether the nucleic acid is single stranded or double stranded and on whether the sequence is contiguous or noncontiguous.

When a palindrome is single stranded and contiguous, a *hairpin* structure can form. One end of the palindrome is complementary to the other, so the nucleotide strand is able to fold back on itself and form base pairs in the region of the palindrome. This is illustrated in Fig. 10-6a. The hairpin region is a double helix even though the nucleic acid is a single strand. This is an important point to keep in mind.

If the palindrome is noncontiguous (i.e., it contains an intervening sequence), then when the strand folds back on itself, the conformation is a *stem-and-loop* structure: a stem where the palindrome bases are self-complementary, and a loop where the intervening sequence is not self-complementary. This is shown in Fig. 10-6b.

When a palindrome sequence occurs in a double-stranded DNA, each of the two strands contains its own palindrome. One palindrome is the complementary sequence of the other, and both palindromes (by definition) are their own complement when read backward. The result is that *both strands* are able to form either a hairpin or stem-and-loop conformation (a hairpin if the palindrome is contiguous or a stem and loop if not). When both strands form a hairpin or a stem and loop, the resulting structure is called a *cruciform*. See Fig. 10-7. Some cruciform structures have been shown experimentally to be binding sites for specific proteins. They also can have a significant influence on tertiary structure of DNA.

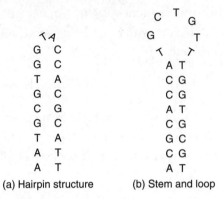

```
                                          C    T
                                       C          G
                       T A          G                T
                       G   C           T          T
                       G   C              A   T
                       T   A              C   G
                       G   C              C   G
                       C   G              A   T
                       G   C              C   G
                       T   A              G   C
                       A   T              C   G
                       A   T              A   T

                  (a) Hairpin structure    (b) Stem and loop
```

FIGURE 10-6 · A single-stranded DNA or RNA palindrome. (a) In a hairpin structure. (b) In a stem-and-loop structure.

```
5'...ATCAGGACTTCAAACCAGGCTAAGAGCCTGGTTTTTTTAGTTCGA...3'

3'...TAGTCCTGAAGTTTGGTCCGATTCTCGGACCAAAAAAATCAAGCT...5'
```

```
                              A
                          A       G
                          T A
                          C G
                          G C
                          G C
                          A T
                          C G
                          C G
                          A T
                          A T
                          A T
                          C     T
    5'...ATCAGGACTT              TTTAGTTCGA...3'
    3'...TAGTCCTGAA              AAATCAAGCT...5'
                          G     A
                          T A
                          T A
                          T A
                          G C
                          G C
                          T A
                          C G
                          C G
                          G C
                          A T
                          T     C
                              T
```

FIGURE 10-7 · A cruciform structure can occur when a double-stranded DNA contains a palindrome.

Still Struggling

Primary structure in DNA is just sequence of nucleotide residues making up the polymer. You just learned about a lot of different secondary structures in DNA. To try to simplify matters, let's briefly review them here. Both single-stranded and double-stranded nucleic acids form a helical secondary structure due to base stacking interactions. When a nucleic acid forms a helix of two strands (double stranded), we call it a *double helix*. Even within the double-helix secondary structure there are various conformations. These include A-DNA, B-DNA, C-DNA, D-DNA, Z-DNA, and others. These differ from each other in terms of tilt of the base pairs, helical pitch, width of the helix, size of the major and minor grooves, helical direction (left-handed vs. right-handed), and so on. DNA-RNA hybrids, where the double helix is one strand of DNA and one strand of RNA, occur briefly during transcription. Evidence suggests that while B-DNA is the most common conformation in nature, DNA-RNA hybrids are most stable in the A-form conformation. In addition to the various helical conformations, local secondary structures also occur, for example, kinks in the helix, bubbles, loops, hairpins, stems and loops, and cruciforms.

DNA Melting

In Chap. 9 you learned that when a protein is unfolded out of its natural conformation, it is said to be *denatured*. *Denaturation* in DNA also refers to the loss of the molecule's typical secondary and tertiary structure. In DNA this most often means unwinding of the helix. But it is not necessarily the case that an unwound DNA helix is out of its *natural* state. In fact, formation of unwound, single-stranded regions in DNA is a natural part of DNA function. The DNA double helix must unwind for nearly all of its biological activities (transcription, replication, etc.). For this reason DNA unwinding is commonly called *melting* (although the term *denaturation* is sometimes still applied).

Denaturation and Helix-Coil Transitions in Nucleic Acids and Proteins Compared

Another term used to describe melting and denaturation in nucleic acids and proteins is the *helix-coil transition*. In this context you should always remember that the term *coil* (in the expression, helix-coil) is short for *random coil*. The melting transition, or helix-coil transition, is from a highly ordered helical state, to a less ordered random-coil state. That is, in the *coil* state, the polymer has a no specific secondary structure. Instead the polymer chain is randomly configured like a piece of string or rope tossed aside.

Proteins typically contain both helical and nonhelical (beta-sheet) secondary structures. So denaturation in proteins includes helix-coil transitions only to the extent that the protein molecule contains helical structures. By contrast, nucleic acid secondary structure is almost exclusively helical, with some exceptions such as kinks and loops. Even cruciform and stem-and-loop formations are structures made up mostly of helix. Therefore denaturation or melting in nucleic acids always involves a helix-coil transition. Also, in *double-stranded* DNA, the transition from helix to coil involves going from a double-stranded to a single-stranded conformation. And formation of single-stranded regions in DNA is a natural part of DNA function. Table 10-2 compares denaturation in DNA and proteins.

TABLE 10-2 Comparison of denaturation in DNA and proteins.		
	DNA	**Proteins**
Common terms for denaturation	Melting, helix–coil transition.	Denaturation, unfolding.
Helix–coil transition	Always.	Only to the extent that the protein contains some alpha–helix segments.
Reversibility	Relatively easy.	Typically not reversible, but may be under specific conditions.
Biological function	Helix–coil transition is an intrinsic part of normal function. DNA must unwind, and later wind back up again, during transcription, replication, and other functions.	Protein folding is the final part of protein construction, in order for a protein to become useful. Some conformational changes take place during biological function, but denaturation is typically destructive rendering the protein useless.

DNA Melting Studies

One of the ways we can unwind DNA is by raising the temperature. Anything that denatures a nucleic acid or protein to is called a *denaturing agent*. Denaturing agents that unwind DNA include temperature, strong acids, strong bases, alcohols, and certain organic compounds such as urea and formamide. All of these are typically used in the laboratory. In cells, there are enzymes that can catalyze the unwinding of DNA. The most common of these is an enzyme called *helicase*. We will see later that other enzymes also affect the unwinding of DNA, but indirectly, through modifying the DNA's tertiary structure. In addition to this, some proteins that bind to DNA specifically stabilize the unwound state, and so contribute to a favorable free-energy change for unwinding the helix.

DNA melting studies are usually done by slowly raising the temperature of a sample of DNA. Alternatively one can gradually add a chemical denaturing agent. The melting transition can be detected in a variety of ways. The two most common are measuring absorbance at a wavelength of 260 nm or measuring the average excess heat capacity. Let's talk about each of these. Nucleotide bases absorb ultraviolet light with an absorption maximum at 260 nm. When the bases are stacked in a helical structure (even single stranded), some of the absorbance is quenched. This is because the electrons that participate in base-stacking interactions are energetically constrained from absorbing photons. As the helix melts, base-stacking interactions are lost and absorbance increases, reaching a maximum when all of the molecules in the sample are completely in the *coil* state. This can be seen in Fig. 10-8.

The halfway point of the melting transition is called the melting temperature and is designated T_m. It is at this temperature where we interpret it to mean that the molecules are, on average, halfway melted. However, depending on

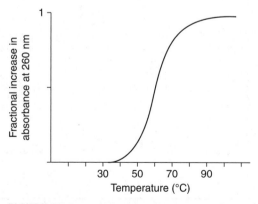

FIGURE 10-8 · Absorbance melting profile for poly-A/poly-T, a synthetic DNA double helix consisting entirely of adenine nucleotides on one strand, and thymine nucleotides on the other.

how energy is distributed among the molecules some may be more than halfway melted, and others less.

The average excess heat capacity is another convenient way to measure the melting transition in DNA. *Heat capacity* is defined as the amount of heat required to raise the temperature of a sample by one degree. At the point where the sample undergoes the melting transition, some of the energy goes into unwinding the helix instead of into raising the temperature. So the heat capacity (the amount of energy required to raise the temperature) appears to increase. At temperatures above the transition, the heat capacity is again similar to what it was before the transition. This is because energy is no longer needed for the melting transition and so all of the energy again goes into simply raising the temperature. Figure 10-9 shows the average excess heat capacity as a function of temperature for poly-A/poly-T. We can measure heat capacity as a function of temperature using an instrument called a *differential scanning calorimeter*. The term *differential* is used because it measures the difference between energy flowing into our sample and energy flowing into a reference sample such as plain water. *Scanning* is used simply because we are scanning over a range of temperatures as we gradually raise the temperature. And a *calorimeter* is of course (see Chap. 2) an instrument for making thermodynamic measurements.

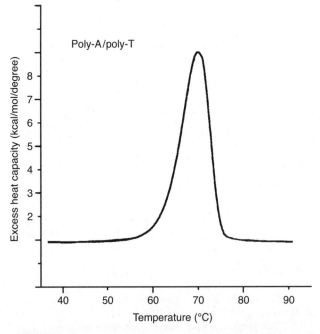

FIGURE 10-9 • Heat capacity melting profile for poly-A/poly-T.

The two typical methods for following the helix-coil transition in DNA have advantages and disadvantages. Absorbance spectroscopy requires only a small amount of DNA, but it does not give direct thermodynamic data except for the melting temperature. Differential scanning calorimetry provides direct thermodynamic measurement of the heat capacity, enthalpy change, and melting temperature. These in turn can be used to calculate the entropy change and Gibbs energy change, but calorimetry requires relatively large amounts of DNA for each experiment.

Synthetic versus Natural Nucleic Acid Melting Studies

By chemically synthesizing DNA (as opposed to extracting natural DNA from a living cell), we can create DNA with a much simpler nucleotide sequence than is found in nature. Simpler DNA is easier to understand, and studying such DNA provides insights that can then be applied to understanding more complex natural DNA. When the nucleotide sequence is simple and repetitive throughout the molecule, we say that the sequence is *homogenous*. An example is synthetic poly-A/poly-T where one nucleotide strand is made of entirely adenine residues and the other strand is entirely thymine residues. Another common laboratory example is poly-AT, where the sequence of both strands consists of alternating adenine and thymine resides, as shown here.

5' ATATATATATATATATATATATATATATAT 3'
3' TATATATATATATATATATATATATATATA 5'

In contrast to these synthetic DNA, the sequence of natural DNA is *heterogenous*. Melting studies of homogenous and heterogenous sequence DNA have shown clearly that the sequence of DNA has a significant effect on how the molecule unwinds.

Model for Helix-Coil Transition in Homogenous DNA

We envision the helix-coil transition in synthetic, homogenous sequence DNA to take place as follows: As the temperature is raised, unwinding begins at multiple locations along the DNA double helix. These locations, where unwinding begins, are more or less random, with one exception: the ends of the molecule tend to melt first. The number of different locations where melting begins can vary from molecule to molecule. In part this depends on the length of the molecule. A longer DNA molecule is more likely to have unwinding occur simultaneously at more locations along the helix, than a shorter DNA molecule.

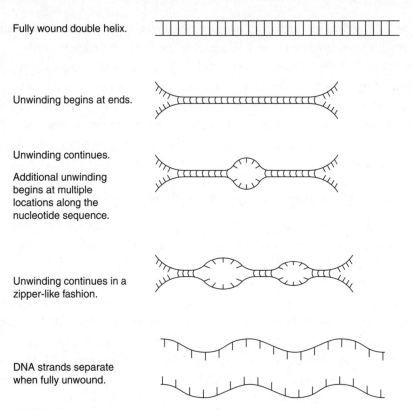

Fully wound double helix.

Unwinding begins at ends.

Unwinding continues.

Additional unwinding begins at multiple locations along the nucleotide sequence.

Unwinding continues in a zipper-like fashion.

DNA strands separate when fully unwound.

FIGURE 10-10 · Model for the helix-coil melting transition in homogenous sequence DNA.

Once unwinding begins it then proceeds in a zipper-like fashion in both directions from each location where it began. (Of course, unwinding that begins at the ends of the molecule obviously can only proceed in one direction: away from the ends and in toward the center of the molecule.) Figure 10-10 illustrates this model of melting in homogenous sequence DNA.

Unwinding at the Ends or in the Middle in Homogenous DNA

Although melting begins at several more-or-less random locations along the helix, the exception to this rule is that the ends of the molecule are most likely to begin melting first. The reason for this is simple. The Gibbs energy change for unwinding a nucleotide base at an end of the molecule is more negative than it is for unwinding a base in the middle of the molecule. We can understand this better by taking a close look at the Gibbs energy change for unwinding the very first few bases (or base pairs) in an otherwise completely helical molecule.

When the entire molecule is in the helical state, each base is constrained within the geometric parameters as noted in Table 10-1. The force holding the bases in their helical conformation comes primarily from base-stacking interactions. In the case of a double helix, hydrogen bonds between base pairs also play a role, but the majority of the energy still comes from the aromatic stacking interaction. Let's say we have a DNA molecule with its bases entirely in their helical conformation. The Gibbs energy change for shifting a single base from its helical conformation to the random-coil conformation has two components. Recall that $\Delta G = \Delta H - T \Delta S$. The first component is the *enthalpy* change ΔH necessary to overcome the stacking interaction. (We can ignore the enthalpy needed to break the base pair hydrogen bonds because its contribution is small compared with the enthalpy of stacking interactions and because we have to break the same number of hydrogen bonds regardless of whether the base is in the middle or at the end of the DNA strand.) The second component is the *entropy* change ΔS for a single base to going from its helical to random-coil conformation. Remember that entropy is a measure of how many different ways there are to achieve a given state or energy level.

For a given helical conformation (as specified in Table 10-1), there is pretty much only one position, one way, for the base to be situated. However, once the stacking interaction is broken, the base is free to swivel about. The covalent bond that attaches the base to the deoxyribose allows the base to freely rotate. Only steric interactions (bumping into other parts of the molecule) limit its motion. This free rotation allows a lot of different positions for the individual base, all of which have the same amount of energy, and all of which are the ways to achieve the random-coil conformation. Therefore the random-coil conformation represents a much *higher entropy* state for the base than when it is constrained into its position within the helical conformation.

Now let's compare the case where the first base to unwind is at the end of the molecule, versus the case where the first base to unwind is in the middle of the molecule. Both components of the Gibbs energy contribute to making it more likely that a base at the end of the molecule will unwind before a base in the middle of the molecule. At the end of the molecule, the last nucleotide base has an adjacent base on only one side. So there is only a single stacking interaction to overcome. At temperatures just below the melting temperature, ΔH is only slightly positive, representing the amount of energy that must be added to the molecule to break the single stacking interaction. However, in the middle of the molecule, there are adjacent bases on both sides of any given base, so there are two stacking interactions to overcome. ΔH for overcoming the stacking interaction is more positive for the base in the middle of the molecule. The base at the end of the molecule, with the smaller positive ΔH, is that much closer to having a negative ΔG and spontaneously unwinding.

The second component of the Gibbs energy, entropy, also makes it more likely that, for a homogenous DNA, the end of the molecule will unwind first. We just saw that the entropy change for a single base going from its helical state to its coil state is positive, since there are more ways to achieve the coil state. A positive entropy makes a favorable contribution to the Gibbs energy change ($\Delta G = \Delta H - T \Delta S$). But a single, coil state base in the middle of the molecule is not as free to move as a single, coil state base at the end of the molecule. In the middle of the molecule, our lone, coil state base is surrounded on both sides by bases that are still constrained in their helical state. This limits the movements of both the sugar phosphate backbone and of the coil base itself. A single, coil state base at the end of the molecule has more ways that it can move and still be in the random-coil conformation for that base. In both cases ΔS is positive. But for the base at the end of the molecule ΔS is larger. Both the larger ΔS and the smaller ΔH for the end base make ΔG more negative for the base at the end of the molecule than it is for the base in the middle of the molecule. Figure 10-11 compares unwinding

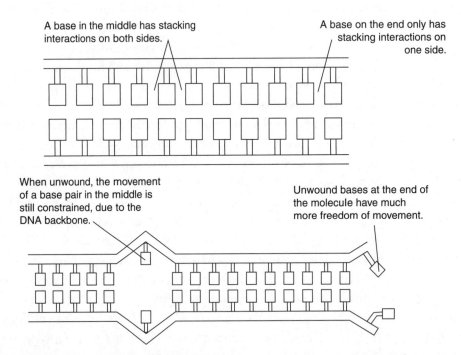

FIGURE 10-11 • Unwinding the very first base.

the first base pair in the middle of the molecule, with unwinding the first base pair at an end of the molecule.

Unwinding Continues

Once we have melted a single base, unwinding an additional base next to the already unwound base is still more likely at the end of the molecule than it is in the middle of the molecule. Regarding bases in the middle, unwinding a second base next to the first is easier than it was to unwind the first base, because we now only have a single stacking interaction to overcome. But it is still less likely than continuing the unwinding at the end of the molecule. Why? The answer in this case involves only the entropy component of the Gibbs energy. It is true that in the middle of the molecule, unwinding the second base increases both its own entropy and the entropy of the first unwound base, since the first base is now also less constrained (since it now has a melted base on one side). However, both bases in the middle of the molecule are still more constrained than two melted bases at the end of the molecule. This is illustrated in Fig. 10-12. The two melted bases in the middle of the molecule are still somewhat limited in their ability to move, because together they are attached on both sides to bases that are still in their helical state. However, at the end of the

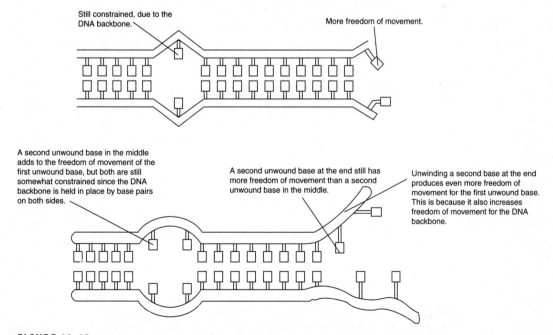

FIGURE 10-12 · Unwinding the second nucleotide base.

molecule, only one of the two unwound bases is attached to a base in its helical state. This gives these two unwound bases a larger degree of freedom than the two unwound bases in the middle of the molecule. In other words ΔS is larger for unwinding the second base at the end of the molecule than for the second base near the beginning of the molecule.

This same logic can be continued for unwinding the third nucleotide base, and the fourth, and so on. However, as we unwind more and more adjacent bases, the loop in the middle of the molecule gets larger and larger. As the loop gets larger, it gains freedom to move about. The constraining effect of having both sides of the loop attached to helical regions of the molecule becomes smaller. It is still not as free to move as an unwound region (of the same size) at the end of the molecule, but the difference between them becomes smaller. Eventually, if we unwind a large enough segment of the molecule, the Gibbs energy change for unwinding the entire segment is approximately equal regardless of whether that segment is in the middle of the molecule or at the end. Unwinding at the end will only be slightly more likely than unwinding at the middle. See Fig. 10-13.

Unzipping DNA

There is another aspect of DNA melting worth looking into. Why does unwinding of additional bases or base pairs typically continue in a zipper-like fashion, from where unwinding began? Why not simply unwind single bases here and there throughout the molecule? The tendency of bases or base pairs to unwind

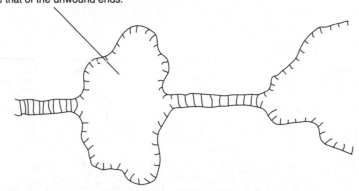

As an unwound portion in the middle becomes large, its freedom of movement becomes more like that of the unwound ends.

FIGURE 10-13 · As an unwound portion in the center of the molecule gets large, the freedom of movement of the unwound loop begins to approximate that of an unwound end.

next to already unwound bases is a type of *cooperativity* (which was briefly discussed in Chap. 2). In fact the tendency is so strong that it is almost impossible to open or unwind just a few base pairs. Instead, entire segments of the helix tend to unwind together very quickly, almost simultaneously. The number of nucleotides that tend, on average, to unwind together is called the *cooperative unit length*. The cooperative unit length can vary depending on the size of the DNA molecule, the sequence and homogeneity of the bases, and various environmental conditions (temperature, salt concentration, etc.). We can explain the cooperative nature of the helix-coil transition in two ways. First, using logic similar to what we used previously, unwinding a base next to an already unwound base takes less energy (enthalpy) compared with unwinding a base that is surrounded on both sides with bases in their helical state. This, as mentioned, is because there is only one aromatic stacking interaction to overcome instead of two. We can also explain this cooperativity in terms of entropy, and within the context of entropy considerations, there are two competing entropy effects, one that favors cooperativity and one that works against it.

The two entropy effects that affect cooperativity are (1) the entropy change due to unwinding an additional base pair as part of an already unwound segment (as opposed to unwinding a lone base pair surround by helical bases) and (2) the entropy of the entire molecule in terms of unwinding long segments of the molecule versus unwinding single bases or smaller segments scattered about along the length of the helix. The first entropy effect was discussed above. Unwinding bases next to an already unwound region increases the size of that region, which in turn increases the mobility of that region. This is so regardless of whether the region is at the end of the molecule or a loop in the middle. This increased freedom of movement means increased entropy, more ways in which the molecule can achieve the same energy state. So unwinding bases next to an already unwound region has a more favorable entropic contribution to the Gibbs energy change, as compared with continuing the unwinding by starting to unwind new regions of the molecule.

Notice that the same logic that told us it's easier to unwind DNA at the ends of the molecule as compared to unwinding in the middle of the molecule also tells us that it's easier to unwind the DNA in a zipper-like fashion, that is, continuing to unwind a region that has already begun to unwind as opposed to starting to unwind an all new segment. One could ask the question why unwinding would even begin at more than one location in the first place. Why doesn't unwinding just begin at the ends and proceed toward the middle until it's done?

The answer lies in considering the entropy change for the entire molecule. We can explain this best by way of an example. Suppose we have a simple, homogenous, double-helical DNA molecule 1000 base pairs long. And let's say this DNA molecule has enough energy to melt 300 base pairs. The question is which contributes more to a favorable free-energy change, melting 300 bases in a row, melting 2 segments of 150 bases each, 3 segments of 100 bases, etc. up to 300 different segments of 1 base pair each?

We know that from an enthalpy point of view, one long segment of 300 base pairs is favored, because each new segment requires a small amount of additional enthalpy to break the extra stacking interaction required to start a new segment. And we know that considering the entropy contribution of each individual base, one long segment is also favored, because again starting a new segment means melting a base with a smaller favorable entropy change compared with melting a base next to an already melted segment. But what about the entropy change for the overall molecule?

To keep the discussion simple, for now, we ignore the small differences in the enthalpy change that result from having multiple melted segments, as compared with melting a single segment. This difference in ΔH is small when we are considering only a few segments, since it is just the addition of one more base-stacking interaction per segment, relative to total of at least 300 stacking interactions. Even if we consider ΔH resulting from melting 300 separate one-base-pair segments, 300 additional base-stacking interactions is relatively small compared to the large difference in ΔS. So for now we will assume that no matter how we arrange 300 unwound bases along a 1000-base-pair molecule the enthalpy level of the molecule is the same (only the entropy is different). With this assumption, we can say that the different ways of arranging 300 unwound bases along a 1000-base-pair molecule represent different ways of achieving the same energy level, and therefore are directly related to the entropy of the molecule. The total number of ways to arrange 300 unwound bases and 700 helical bases along a 1000-base-pair molecule is given by

$$W = \frac{1000!}{(300!) \cdot (700!)} \tag{10-1}$$

This is approximately equal to 5.428×10^{263}, a rather large number of ways to achieve the same state of 300 unwound base pairs. But this large number includes all possible distributions of the 300 unwound base pairs, from those with a single 300-base-pair unwound segment, through those with 300 separate unwound segments. In practice not all of these distributions of unwound bases are likely to occur. Rather there will be some average cooperative unit length somewhere between a single base pair (not cooperative at all) and 300 base pairs. Or, in terms of the number of melted segments, somewhere between 300 melted segments (of one base pair each) and a single segment (of 300 base pairs)

In order to consider the entropy differences for different numbers of segments, we need to calculate the number of ways of arranging the unwound bases separately for each possible distribution of unwound segments. The mathematics gets a little beyond our scope, so we'll just walk through the first few cases to give an idea where it's leading.

Take the case of a single segment of 300 unwound base pairs. How many ways are there to arrange a single segment of 300 unwound base pairs in a 1000-base-pair molecule? The answer is illustrated in Fig. 10-14. There are 701 possible ways to arrange 300 *contiguous* unwound bases within a 1000-base-pair DNA. Therefore the case of a single 300 base pair melted segment represents only 701 of the 5.428×10^{263} total number of ways to arrange 300 melted base pairs with any number of segments.

Now consider the case of two segments of unwound bases, each 150 base pairs in length. This is illustrated in Fig. 10-15. Let's count the number of ways to arrange two

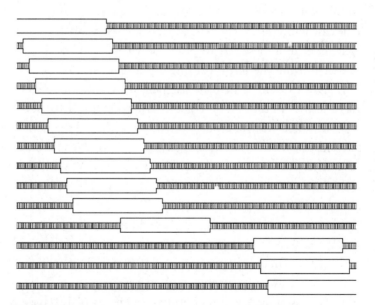

FIGURE 10-14 · Illustration of a 1000-base-pair DNA with 300 contiguously unwound bases. This leaves 700 base pairs still in their helical state. The 300-base-pair segment can be at the end of the molecule or anywhere in the middle with various numbers of the 700 helical base pairs on either side. The illustration shows 14 of the 701 possible ways to have a single unwound segment of 300 base pairs.

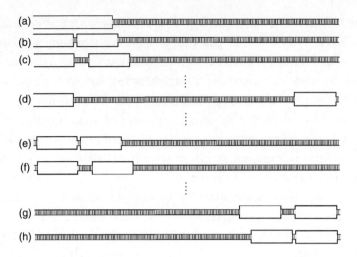

FIGURE 10-15 · Illustration of a 1000-base-pair DNA with two segments of 150 contiguously unwound bases each.

such segments. If we keep the left-most unwound segment at the left end of the molecule, the other segment (of 150 unwound base pairs) can be shifted one base pair at a time along the molecule. For this 150-base-pair segment, the situation is identical to the case illustrated in Fig. 10-14. There are 701 ways to arrange this 150-base-pair segment. Four of these 701 ways are shown as a, b, c, and d in Fig 10-15.

Now consider moving the left-most unwound segment just one base pair to the right. This gives us a single helical base pair at the left end of the molecule (see drawings e and f in Fig. 10-15). For this arrangement of the left-most segment, there are now 700 ways to arrange the other unwound segment. If we then move the left-most segment again by one base pair, there are now only 699 ways to arrange the other (the right-most) unwound segment. We can continue this logic to find that the total number of ways to arrange two 150-base-pair unwound segments along a 1000-base-pair molecule is 701 + 700 + 699 + . . . + 1, or 246,051. This is an amazing result! By simply dividing the 300-base-pair segment into two, we've increased the number of ways to arrange the unwound base pairs from 701 to 246,051. That's a factor of over 350 times! This tells us that based only on the number of ways to arrange melted base pairs, it is 350 times more likely that the 300 based pairs will melt in two segments of 150 base pairs each than it is that they will melt as a single segment of 300 base pairs.

Now consider three segments of 100 base pairs each. If we keep the left-most 100-base-pair segment fixed at the left end of the molecule, the logic for the next two segments (the middle and the right-most segment) is the same as for the previous case of two segments, which gave us 246,051 ways to arrange just those two segments. Now if we move the left-most segment just one base pair to the right and repeat the process for the other two 100-base-pair segments, we get $700 + 699 + \ldots + 1$, or 245,350 ways to arrange the two right-most segments. Move the left-most segment again, and we get $699 + 698 + \ldots + 1$, or 244,650. If we continue this logic, we find that there are 57,657,951 ways to arrange three contiguous unwound segments of 100 base pairs each.

In going from just one segment to three segments, we have increased the number of ways to arrange the unwound bases from less than a thousand to more than 57 million! You can just imagine what increasing the number of segments to 4 or 8 or 20 would do. And we have not even yet considered having different numbers of base pairs in each of the segments. For example, take the case of two segments. We considered 150 base pairs in each segment. What about the case of 149 base pairs in one segment, and 151 base pairs in the other? And so on. This means that the case of two segments does not really give us 246,051 ways to arrange the segments, but 299 times 246,051 ways, or 73,569,249 ways. Similarly the case of three segments was undercounted. If we include various lengths for the three segments (e.g., instead of 100, 100, 100, consider 99, 100, 101, and 98, 100, 102, and 98, 99, 103, etc.) the result is more than 2.5×10^{12} ways.

Still Struggling

Where does the 299 come from? This is the number of ways to split 300 unwound base pairs into two segments. The smallest number of base pairs that can be in a segment is one. So we start counting with a segment of 1 and a segment of 299. From there we go to 2 and 298, followed by 3 and 297, and so on, until we get to 299 and 1. But isn't 1 and 299 the same as 299 and 1? The answer is yes and no. Yes they are the same if order does *not* matter. No, they are not the same, if order *does* matter. In a homogenous sequence DNA order does not matter. So for the case of homogenous sequence DNA we are double counting. But most DNA sequences are heterogenous. Certainly natural DNA are, so for most DNA we are not double counting. The important point to realize here is that splitting the unwound base pairs up into multiple segments so greatly increases the number of ways of arranging them along the molecule, that it doesn't make a significant difference whether we are double counting (the additional factor of how to split up the base pairs among the segments) or not.

It's easy to see from this, that unwinding more, smaller segments results in a much larger positive entropy change than unwinding just a few or one contiguous segment. So when we look at the entropy change for the whole molecule, having more segments is more favorable to the Gibbs energy change than having just a single segment. However, entropy considerations at the individual base level have the opposite effect. These two opposing effects balance each other out. The end result is that DNA unwinds in multiple segments, of some average cooperative unit length. To summarize, there are two competing entropy effects. They are (1) the entropy of coil state base pairs related to their freedom of movement favoring longer contiguous unwound segments and (2) the entropy due to the number of ways of arranging unwound bases along the molecule. This latter entropy favors more, shorter, unwound segments. It is possible to calculate the balance between these two effects, after making some assumptions about the enthalpy change, to arrive at a theoretical cooperative unit length which can be compared to experimental observations.

The Assumption of the Same Energy State

Now let's deal with the assumption that the energy of the molecule is the same regardless of how we arrange the unwound bases. One of the ways we can deal with this assumption is by melting fewer base pairs when we need extra enthalpy to break the extra stacking interactions. If we do this, we are not so much removing the assumption but actually strengthening the basis for it. The technique is this: We hold the energy of the molecule constant by melting fewer base pairs in the case of those arrangements that would otherwise require more energy to break the additional stacking interactions. By melting fewer base pairs we can keep the energy constant. In this way we *can* assume that the energy of the molecule is constant for all of the different arrangements, with no doubts about that assumption. If so, then all of the various arrangements (for a given energy level) truly represent different ways of achieving the same energy state of the molecule, and therefore truly represent the entropy state of the molecule.

So, for example, if the molecule has enough enthalpy to overcome 300 stacking interactions, we can unwind 300 contiguous bases at one end of the molecule. If, as another arrangement, we unwind ten separate segments (instead of one), then we have ten additional stacking interactions to overcome. Given the same amount of enthalpy, we can only unwind 290 base pairs. So we would unwind 10 segments of 29 base pairs each (instead of 10 segments of 30 base pairs each). The same argument applies as we increase the number of segments. In the extreme case we can melt a total 150 base pairs in 150 separate segments.

This is because 150 separate segments (not at the ends of the molecule) require disrupting 300 base-stacking interactions (one on each side of each segment). In this way we can safely assume that the energy of the molecule is constant for all of the different arrangements. The only difference between this analysis and our previous analysis is that we are now considering a somewhat different set of arrangements because the arrangements have different numbers of melted base pairs. It is still very much the case, however, that as we increase the number of segments, then the number of ways of arranging them increases tremendously, as we saw above.

The Helix-Coil Transition in Heterogenous DNA

Figure 10-16 shows the melting profile for a heterogenous sequence DNA. The DNA was derived from naturally occurring DNA. There is no particular or obvious pattern to the sequence of nucleotides. Some of the sequences are actual genes that contain the code for building specific proteins. A portion of the sequence is shown in Fig. 10-17. Notice that the melting profile contains multiple peaks, in contrast to the single-peak melting profile of an homogenous DNA in Fig. 10-9. This is indicative of the fact that different portions of the nucleotide sequence are easier to melt than others. The portions that are easier to melt do so at lower temperatures, whereas other portions unwind at higher temperatures.

As we mentioned toward the beginning of this chapter, the stacking interactions between various nucleotide bases are not all equal. We gave the example

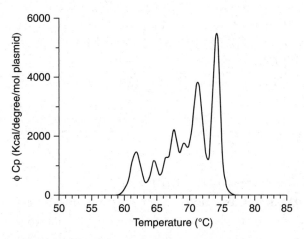

FIGURE 10-16 · Heat capacity melting profile for pBR322 DNA.

```
TTCTCATGTT  TGACAGCTTA  TCATCGATAA  GCTTTAATGC  GGTAGTTTAT  CACAGTTAAA
TTGCTAACGC  AGTCAGGCAC  CGTGTATGAA  ATCTAACAAT  GCGCTCATCG  TCATCCTCGG
CACCGTCACC  CTGGATGCTG  TAGGCATAGG  CTTGGTTATG  CCGGTACTGC  CGGGCCTCTT
GCGGGATATC  GTCCATTCCG  ACAGCATCGC  CAGTCACTAT  GGCGTGCTGC  TAGCGCTATA
TGCGTTGATG  CAATTTCTAT  GCGCACCCGT  TCTCGGAGCA  CTGTCCGACC  GCTTTGGCCG
```

FIGURE 10-17 · The first 300 nucleotides in the sequence for pBR322 DNA.

that the free-energy change of stacking an adenine on top of a thymine is –0.78 kcal/mol, whereas the free-energy change from stacking adenine on top of guanine is about –1.19 kcal/mol. Both are negative and so both are stabilizing interactions, but clearly the A on top of G interaction is more stabilizing than A on top of T. It takes more energy to disrupt an AG stack than and AT stack. Table 10-3 lists free-energy changes for various combinations of stacked nucleotides. You can see clearly from the table that stacking energies that involve guanine and cytosine are significantly higher than those that involve adenine and thymine.

TABLE 10-3	Gibbs energy changes at 37 °C for base stacking various combinations of nucleotides.
Stack	**Energy (kcal/mol)**
A/T	−0.78
T/A	−0.65
T/G	−1.39
A/C	−1.33
A/G	−1.19
T/C	−1.32
C/G	−2.19
G/C	−2.19
C/C	−1.82
T/T	−0.98

You can see that there is quite a variation in the Gibbs energy change required to disrupt a stacking interaction (in order to unwind a base) depending on which base is next to it. This means that melting of heterogeneous sequence DNA does not necessarily begin at the ends of the molecule, nor does it begin at more-or-less random locations along the helix. Rather, as we add energy to the system, melting begins where the Gibbs energy of stacking is the smallest. Of course, all of the considerations mentioned for homogenous DNA have to be taken into account. This includes the enthalpy change and the entropy due to freedom of movement for each base, as well as the entropy associated with the number of ways of arranging the unwound bases. But we now also must consider the variation in stacking interactions, which depend on sequence. These variations may limit the possible arrangements of helix and coil segments. For a given energy level, not all arrangements are equally likely. Instead the arrangements will be limited to those that involve unwinding segments which are easier to unwind.

As we saw, the entropy associated with the number of ways to arrange unwound bases favors having more segments with shorter cooperative unit lengths. So limiting the number of ways to arrange the melted base pairs can have the opposite effect, increasing the cooperative unit length and melting fewer segments. On the other hand, the sequence itself can also have an effect on the cooperative unit length. For example, if a section of helix contains a lot of easily melted bases, and that section is surrounded by sequences that are difficult to unwind, then the easy melting sequences will clearly melt first. Under these conditions, the length of the easy melting region has a strong influence on the cooperative unit length.

The end result is that the sequence of nucleotides not only carries the genetic code, but can have a significant influence on which regions of the double helix are easy to unwind and which regions are difficult. This occurs both through the strength of specific stacking interactions, as well as through their influence on the configurational entropy of the unwound bases. This effect is important biologically because DNA must unwind at specific locations in order to perform many of its biological functions, such as transcription and replication.

The Nearest-Neighbor Approximation

We can calculate the effect of sequence on unwinding DNA. Although the calculations are beyond the scope of this book, it's worth discussing them here to gain an understanding of how we do it. One thing is clear; regions that are rich in adenine and thymine melt easier than regions that are rich in guanine and cytosine. The free energy required to disrupt a GC, GG, or CC stack can be as much as 2 to 3 times that needed to disrupt an AT, AA, or TT stack. Although this free energy is influenced by the sequence of bases several base pairs away, by the number of hydrogen bonds, and by steric interactions, we can make an approximation in our calculations. The approximation is called the *nearest-neighbor approximation*, which means that when accounting for sequence in a heterogenous DNA, we only consider the base stacking interactions of bases that are immediately next to each other. The nearest-neighbor approximation gives results that agree very well with experiment; this suggests that factors other than the nearest-neighbor stacking energies play a minimal role in the effect of sequence on DNA melting.

The Two-State Approximation

Another approximation we can make when analyzing a melting profile for heterogeneous DNA is to assume that each segment unwinds in an all-or-none manner (rather than the zipper-like conformational transition we have been describing up until now). The partition function for a two-state transition is given by

$$Z = 1 + e^{-\frac{\Delta H}{RT} + \frac{\Delta S}{R}}$$ (10-2)

This is similar to Eq. (5-16), except that the Boltzmann constant k has been replaced by the gas constant R to reflect the convention to report thermodynamic values (such as the enthalpy and entropy) in units per mole of base pairs. The other difference with Eq. (5-16) is that there is only a single exponential term instead of the sum on the right side of the equation. This is because we are only dealing with two states. ΔH and ΔS are the enthalpy and entropy change between state A and state B, where state A is the fully wound helix, and state B contains the fully unwound *segment* of n_c base pairs, where n_c is the cooperative unit length—the number of base pairs that melt together in an all-or-none manner (thus giving rise to the appearance of a two-state transition).

If we calculate the heat capacity melting curve from this partition function, the result is a single bell–shaped curve similar to any one of the peaks in Fig. 10-16. Clearly the overall melting of a heterogeneous DNA is not a two-state transition. But the question of interest is whether each peak in the melting profile represents the melting in an all-or-none manner of a single segment. By definition, the heat capacity is the amount of energy required to raise the temperature of our sample by 1°. The apparent heat capacity increases (as in Fig. 10-16) while the DNA is undergoing transition from helix to coil since some of the energy is going into unwinding the DNA (instead of all of it going into raising the temperature of the solution). This energy is the enthalpy of the melting transition. If we multiply the heat capacity (kilocalories per degree) times the range of temperature (degrees), we get the enthalpy for the transition. In other words, the area under the heat capacity curve is the enthalpy change for the melting transition. To see if each peak in the melting profile represents a two-state transition, we fit each peak in the experimental melting curve with a heat capacity curve calculated from the following partition function:

$$Z = 1 + \omega e^{-\frac{\Delta H}{RT} + \frac{\Delta S}{R}}$$

(10-3)

Notice that this is very similar to the partition of Eq. (10-2), except for the introduction of the variable ω, which is the number of identical two-state transitions occurring at the same time. For any given peak in the melting profile, if ω is calculated to be equal to (or close to) 1, then ΔH in the exponent will be equal to the area under the peak and we can say with confidence that the peak represents the unwinding of a single segment that unwinds in pretty much an all-or-none fashion. We can calculate the average enthalpy change per mole of base pairs by taking the overall enthalpy change for melting the entire DNA and dividing by the number of base pairs. Let's call that Δh. The ratio of the ΔH for one segment to the average enthalpy per base pair Δh gives us a measure of the cooperative unit length n_c. This measure of cooperative unit length however is skewed for a heterogeneous DNA, because the average per base-pair value Δh was calculated for the entire DNA molecule, whereas the peaks in the melting profile at the lower temperatures correspond to regions of the DNA that are rich in adenine and thymine, and tend to have a smaller enthalpy change. The peaks at higher temperatures correspond to regions of the DNA with a higher percentage of guanine and cytosine. So n_c calculated from an average enthalpy for the overall DNA will tend to be underestimated for the lower temperature peaks and overestimated for the higher temperature peaks. However, using nearest-neighbor interactions and knowing the DNA sequence, we can calculate individual average per-base-pair enthalpies for each region of the melting profile and, in this way, refine our estimates of the cooperative unit length n_c for each portion of the melting profile.

If, for a given peak in the profile, ω turns out to be significantly greater than 1, then we know it is unlikely that the DNA segments in that portion of the melting curve are unwinding in a two-state fashion. In such cases a more gradual zipper-like model or a model that involves multiple two-state transitions (unwinding of multiple segments each in an all-or-none manner) is closer to reality. In the latter model ω represents the number of segments of DNA melting in a two-state manner and n_c would be the average length of each segment under that peak.

DNA Tertiary Structure

The DNA double helix is not a rigid rod. The molecule is somewhat flexible, not quite as flexible as a piece of thread or string but somewhat flexible like a garden hose. The double helix itself it is able to bend and twist. Flexibility, or rather stiffness, in polymers is often measured using persistence length. The *persistence length* is the length of a segment of polymer that behaves as if it is relatively rigid. Beyond the persistence length, the polymer can bend, so, the shorter the persistence length, the more flexible the polymer.

Flexibility in DNA

One way to mathematically model the flexibility of a polymer is to consider it to be a series of rigid segments that are freely jointed together. This means that each segment is free to angle and rotate relative to its adjacent neighboring segments. The length of each segment is the persistence length of the polymer. In reality the molecule may not be a chain of freely jointed rigid segments, but this model often works well to predict properties of polymers that result from flexibility. In this way we can mathematically describe the flexibility of DNA and other biopolymers without the need to measure flexibility directly.

The freely jointed chain model allows us to calculate the *root mean square end-to-end distance*. Each segment is treated as a vector pointing to the next segment. The sum of all of these vectors is a single vector pointing from one end of the chain to the other. Figure 10-18 illustrates this. We can calculate this single end-to-end vector for all possible conformations of the freely jointed chain. We would like to get some average value for this end-to-end vector, as a way of characterizing the set of all possible conformations of the chain.

If we consider all possible conformations, the set of all possible end-to-end vectors will include vectors in all different directions. Adding up these vectors will result in some of the vectors canceling each other out (because they point in opposite directions). The average vector will then be zero. We solve this problem by squaring the end-to-end vector. Multiplying vectors together converts them to scalars, so each squared vector is a scalar. We then sum all of the squared end-to-end vectors and take the mean (average) of the squares. The square root converts the units from distance squared back into a simple distance. This *root mean square (RMS) end-to-end distance* gives us one measure of the compactness of the polymer. It turns out that for a freely jointed chain of N segments, each of length L, the *RMS distance* is equal to the square root of the number of segments times the length of each segment.

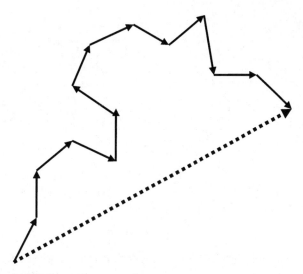

FIGURE 10-18 • Freely jointed chain represented by a series of vectors. The sum of these vectors is a single vector pointing from one end of the chain to the other.

$$\text{RMS end-to-end distance} = \sqrt{\langle r^2 \rangle} = \sqrt{N}L \qquad (10\text{-}4)$$

Superhelical DNA

The most common DNA tertiary structure found in nature is a *superhelix*. The double helix itself bends and curves such that the *axis* of the double helix also forms a helix, as shown in Fig. 10-19. Superhelical DNA is sometimes referred to as *supercoiled* DNA. However, be careful not to confuse the use of coil in *supercoil* with the use coil in *helix-coil transition*. In the case of supercoil, we mean a helix, as in superhelix. But in the case of a helix-coil transition, coil is short for *random coil*. meaning the *opposite* of a helix. For this reason we will generally try to avoid using *supercoil*, and stay with *superhelix*. But be aware that many books and articles will use the word *supercoil*.

You can gain a lot of insight regarding superhelicity by taking a piece of garden hose, or rubber tubing, grabbing it in two places (one with each hand) some distance apart, and twisting it with your hands. The axis of the garden hose will kink up, forming a helical shape. You can take the analogy further by painting (or imagining) a double helix on the garden hose or rubber tubing. Then the kinked up tube or hose truly is a superhelix relative to the double helix that's painted on the hose. Now here's where the insight comes in. As you are holding

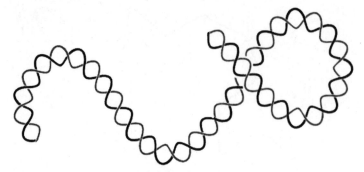

FIGURE 10-19 • The DNA double helix can bend and curve its axis into a helical shape. This helix of a helix is called a *superhelix*.

the kinked-up hose, let go with one hand. What happens? The superhelix goes away. The hose relaxes. The same thing is true of DNA. Unless the ends of the DNA molecule are somehow held in place, the superhelical tertiary structure goes away. Regarding the secondary structure (the double helix) the DNA molecule has the necessary forces internally (base stacking, hydrogen bonds) to maintain its secondary structure, but it cannot maintain the tertiary structure of the superhelix without being somehow held in place. The tertiary structure comes from the fact that the DNA double helix is partially flexible, but with some elasticity, like a garden hose. This partial flexibility and elasticity come from the base stacking, hydrogen bonds, and covalent bonds that hold the double helix together. But, due to the elasticity, it requires some stress on the molecule in order to maintain the tertiary windings.

There are three ways that nature typically holds the ends of the DNA molecule in place in order to maintain superhelical windings. See Fig. 10-20. The first is by binding the ends of the DNA or a portion of the DNA to a protein, thus forming a loop of double helix. The second is simply by winding the double helix around a protein complex. This is the most common way to maintain superhelicity in eukaryotes. The protein complex with DNA wrapped around it is called a *nucleosome*. The third way to hold the double helix in place to maintain superhelicity is for the two ends of the double helix to be covalently attached to one another forming a circular molecule. Circular DNA molecules are most common in bacteria and other prokaryotes.

Once the ends of the double helix are held firm, if the double helix is twisted (or untwisted) to an extent that is different from its normal amount of helical twist, the stress of this twisting will cause the axis of the double helix to kink up. This kinking up is the superhelix. We can best understand the nature of

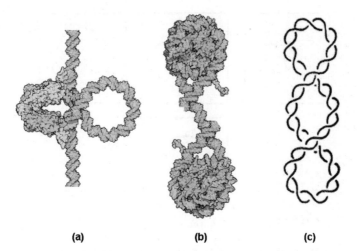

(a)   (b)   (c)

FIGURE 10-20 · Three ways to hold DNA double helix into a superhelical conformation. (a) The ends of a portion of double helix are held in place by a protein. (b) The double helix is wound around a protein complex called a histone. (c) The two ends of the double helix are covalently attached to one another, forming a circular molecule.

superhelix formation by studying circular DNA. This is the third case (c) in Fig. 10-20. Circular DNA is uncomplicated by the need for bound proteins to maintain superhelicity. The principles and the relationship between the double helix and the superhelix are the same in any case. But the ability to examine these principles with regard to DNA alone (without bound proteins) simplifies our understanding.

Geometry and Topology of Superhelical DNA

The branch of geometry and mathematics that deals with objects being bent or deformed in a continuous manner is called *topology*. The bending of the double helix to form a superhelix therefore falls into this area of mathematics. In a circular, double-helical DNA, the two ends of the double helix are covalently bonded together forming a double-helical circle. This type of DNA is called *closed duplex DNA, closed* because the DNA strands each form a closed and continuous circle and *duplex* because there are two strands of DNA. [The term *circular DNA* can also refer to a single-stranded circular DNA molecule; therefore to be more exact, we use the term *closed duplex DNA* (*cdDNA*)]. Closed duplex DNA is found in nature in many organisms and can also be synthesized in the laboratory.

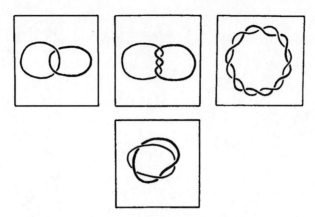

FIGURE 10-21 · The linking number *Lk* is a topological invariant of two or more closed curves. The drawing shows, from left to right, two closed curves that are linked in space one, two, and six times respectively. The center pair of drawings is intended to illustrate that, no matter how the curves are bent or twisted (continuous deformations of shape) the number of times that the two curves are linked together remains constant. The linking number can only be changed by breaking and resealing one or both curves in at least one location along the curve.

Each of the two strands of a cdDNA are themselves continuous, closed circles. But because of the intertwining of the double helix, the two strands are linked to one another like two links in a chain. Figure 10-21 illustrates two closed curves as a way of representing the two circular strands of DNA in a cdDNA molecule. The figure shows the two curves linked together one, two, and six times. Although the links of a chain are typically linked together only once, two intertwining, circular strands can wrap around each more than once. The number of times that one strand is linked to the other strand is called the *linking number* and is denoted by *Lk*. The linking number is a *topological invariant*; it does not change under continuous changes of shape. Only if one or both of the DNA strands is broken (a *non*continuous, or discontinuous, change of shape) can the linking number change.

In the field of topology, regarding linked curves, there are two more parameters that are directly useful for studying and understanding closed duplex DNA. The first is *twist*, denoted by *Tw*. The *twist* is the number of times one curve wraps around the other along the entire length of the curves. For the curves shown in Fig. 10-21, the twist is exactly equal to the linking number. Whenever the twist is equal to the linking number, then the two curves (except to the extent that they twist around each other) will lie entirely within a plane or, more precisely, entirely on the surface of a sphere. This is the case in Fig. 10-21; the curves as shown (except where they twist around each other) lie entirely within a plane, or more precisely entirely on the surface of a large sphere. (*Note:* The sphere can be large or small, whichever is necessary. The point is that if Tw is equal to Lk, then it will always be geometrically possible to define a sphere such that the curves lie entirely on the surface of that sphere. This is the correct way to describe the case where Tw equals Lk. Regarding the idea of lying within a plane, this is used for convenience because it is often simpler to visualize the curves lying within a plane than on a surface of a sphere; the concept of lying within a plane works just as well because a plane is effectively the surface of an infinitely large sphere.)

If the twist is not equal to the linking number, then it is not possible for the curves to lie on the surface of a sphere or plane. We will not prove this topological theorem, but take it as a given. Those of you who are interested can certainly look it up in any decent book on topology or on the Web. The curves, instead of lying in a plane, will bend, or kink, rising above or below the plane. The extent to which the curves are not able to lie in a plane or on the surface of a sphere (but rise above or below it) is another topological parameter called *writhe*, and denoted by Wr. There is a simple mathematical relationship between the linking number, the twist, and the writhe, given by

$$Wr = Lk - Tw \tag{10-5}$$

As we saw in Table 10-1, depending on the particular conformation of DNA, the various forces (base stacking, hydrogen bonds, etc.) stabilize the double helix to have a particular amount of helical pitch. In Table 10-1 this parameter was expressed in terms of the length in angstroms of a single turn of the helix. In a circular DNA there is obviously a relationship between the pitch of the helix and the twist expressed in terms of the number of times one strand winds around the other for the entire circumference of the circle. For a given cdDNA, the twist will be the circumference of the circular DNA divided by the pitch of the helix.

$$Tw = \frac{\text{circumference}}{\text{helical pitch}} \tag{10-6}$$

The DNA double helix has a preferred (most favorable, most stable) helical pitch, which is defined as the point where the Gibbs energy of helix formation is at a minimum (is most negative). The facts that there is a most stable helical pitch and that for a circular DNA there is a direct relationship between the helical pitch and the twist [Eq. (10-6)] means that there is a most stable amount of

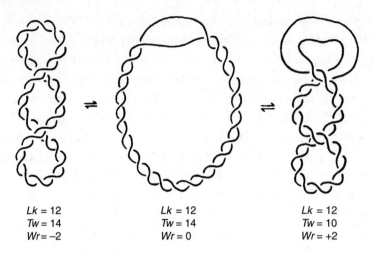

$$Lk = 12 \qquad\qquad Lk = 12 \qquad\qquad Lk = 12$$
$$Tw = 14 \qquad\qquad Tw = 14 \qquad\qquad Tw = 10$$
$$Wr = -2 \qquad\qquad Wr = 0 \qquad\qquad Wr = +2$$

FIGURE 10-22 · Representation of a cdDNA molecule with $Lk = 12$ and different amounts of twist (Tw).

twist. If the most stable amount of twist differs from the linking number, then the DNA will writhe, according to Eq. (10-5). In other words the axis of the double helix will not lie within a plane, but will curve and bend forming a super-helix. Figure 10-22 illustrates this. On the left side of the figure is a cdDNA in which the most stable amount of twist is 14, but the linking number is only 12. Therefore the cdDNA writhes, forming supercoils. In this state the DNA is said to be *negatively supercoiled*, because the writhe is negative.

In the middle of Fig. 10-22 we have unwound some of the base pairs (per-haps by raising the temperature). Assuming 10 base pairs per turn of the double helix, if we melt 20 base pairs, then we unwind two turns of the double helix. Removing two turns of the double helix reduces the twist from 14 to 12. Now the twist is equal to the linking number, the writhe is equal to zero, and the axis of the cdDNA lies on the surface of a sphere or within a plane. This is an impor-tant result. It means that unwinding negatively superhelical DNA reduces superhelicity. Superhelical DNA found in nature is almost always found to be negatively supercoiled.

In the final drawing on the right side of Fig. 10-22, we have melted additional 20 base pairs, bringing the value of Tw down to 10. Now the twist is *less* than the linking number. This forces the molecule to have a positive writhe, and the cdDNA now becomes superhelical in the positive direction. This is also a very important result. It means that if we unwind enough base pairs, then the DNA eventually becomes superhelical with a positive writhe. This is extremely important, because transcription and replication involve unwinding a lot of base pairs. As we unwind more and more base pairs to transcribe or replicate the DNA, positive supercoils form in front of the replication (or transcription) fork. As we unwind more and more double helix, more and more superhelix forms in the still wound portions of the molecule. Bending the axis of the double helix requires energy, and if we bend it enough it could eventually break or at least encounter steric hindrances. The cell has to somehow reduce this positive supercoiling, and it does so with the aid of enzymes called *topoi-somerases*. Topoisomerases are enzymes that change the level of supercoiling in DNA by changing the linking number *Lk*. Individual DNA molecules that are identical except for *Lk* are called topoisomers (short for topological isomers).

Nicked Duplex DNA

What happens if we break one of the strands in a closed duplex DNA? If we break one of the strands of a cdDNA, then the strands are free to turn around each other, and there is no longer a constraint that the linking number must remain constant. In fact, strictly speaking *Lk* is no longer defined. The ends of the double helix are also no longer held in place, which (as noted previously) is necessary to maintain a superhelix. What happens, as illustrated in Fig. 10-23, is that the axis of the helix is free to relax; the writhe goes to zero. A circular,

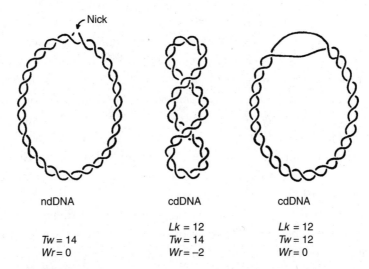

nddDNA

Tw = 14
Wr = 0

cdDNA

Lk = 12
Tw = 14
Wr = −2

cdDNA

Lk = 12
Tw = 12
Wr = 0

FIGURE 10-23 • Representation of a ndDNA molecule with *Tw* = 14, and a cdDNA molecule with *Lk* = 12 and different amounts of twist (*Tw*). When a cdDNA is nicked, the force holding the supercoils in place is removed, and the molecule relaxes until the writhe equals zero.

double-helical DNA with one or more breaks in one or more strands is called *nicked duplex DNA* (*ndDNA*). If both strands are broken, as long as the breaks are not directly across from one another and conditions are favorable for a double helix to form, then the molecule will remain circular.

The fact that a ndDNA will spontaneously relax (the writhe will spontaneously go to zero) is indicative of the fact that there is energy bound up in the superhelix. Once we remove the structural constraint of holding the ends of the double helix in place (by being covalently attached to one another), then the molecule is free to seek a lower Gibbs energy value. In the case of a negatively superhelical *closed* duplex DNA, unwinding the double helix will initially reduce the writhe (as we saw in Fig. 10-22). The energy of that writhe can contribute to unwinding the double helix. This means that unwinding base pairs in a negatively superhelical DNA is easier (requires less energy from the outside) than it is in a linear or ndDNA. The opposite is true for positive superhelical DNA. Unwinding base pairs in positively supercoiled DNA increases superhelicity; this makes it more difficult to unwind the DNA further.

Topoisomerases

As we mentioned briefly above, the cell contains enzymes capable of altering the linking number of DNA. These enzymes are called *topoisomerases*. Topoisomerases alter the linking number by temporarily breaking one or both strands of DNA. (Recall that *Lk* is a topological invariant; it cannot be changed by continuous deformations of shape but only by breaking one or both strands of DNA.) There are two classes of topoisomerases. *Type I topoisomerases* change the linking number by breaking only one strand of DNA, then allowing or forcing one of the strands to wind or unwind around the other, and then resealing the covalent bond that was broken. *Type II topoisomerases* break both strands of DNA, and then pass a section of unbroken double helix through the break before resealing the covalent bonds. In this way type II topoisomerases always change the linking number two at a time, for example, from 14 to 16 to 18, and so on. Type I topoisomerases can change the linking number only one at a time, for example, from 14 to 15 to 16, and so on.

Most topoisomerases reduce the amount of writhe, regardless of whether the writhe is positive or negative. These topoisomerases effectively allow the DNA to relax while one or both strands of the molecule are temporarily broken. This releases some of the energy that is in the writhing. Some topoisomerases, known as *gyrases*, are able to increase negative writhe by reducing the linking number. Increasing the absolute value of writhe requires energy (to bend the DNA into a superhelix). This energy comes from ATP. Gyrases couple their topoisomerase reaction with cleaving one of the high-energy phosphate bonds in ATP and utilizing that energy to increase the writhe in DNA. In this way the cell can store energy in the elasticity of the DNA (like winding up a spring), which can later be used to help unwind the DNA for transcription or replication. It can also, to some extent, control the onset of transcription or replication, by altering the superhelicity; this in turn affects how easy or difficult it is to unwind a particular region of DNA.

QUIZ

Refer to the text in this chapter if necessary. Answers are in the back of the book.

1. What do A-, B-, C-, and Z-DNA describe?

   A. The four nucleotide bases found in DNA.

   B. Four DNA secondary structures: hairpins, loops, cruciforms, and tails.

   C. Double-helical structures that differ in pitch and width of the helix.

   D. The four types of single-stranded DNA.

2. Which statements are true?

   S1. The DNA double helix must unwind for transcription and replication.

   S2. Whether the DNA will unwind depends on the Gibbs energy change.

   S3. Unwinding circular DNA will change the writhe of the helix.

   A. S1 and S2

   B. S2 and S3

   C. S1 and S3

   D. S1, S2, and S3

3. What is a nucleic acid palindrome?

   A. A nucleotide sequence that can form a stem-and-loop or cruciform structure.

   B. A nucleotide sequence that reads the same forward or backward.

   C. A protein that binds to certain DNA sequences.

   D. A curved, dome-shaped secondary structure in nucleic acids.

4. What is homogenous DNA?

   A. DNA that has been blended together.

   B. DNA from only one species.

   C. DNA with a sequence made from only a single nucleotide.

   D. DNA with a simple sequence that repeats throughout the molecule.

5. The root mean square end-to-end distance of a freely jointed chain is proportional to what?

   A. The length of a segment and the square root of the number of segments.

   B. The square root of the length of a chain segment.

   C. The number of segments and the square root of the length of each segment.

   D. The difference between the length of one end of the chain and the other.

6. What is the primary driving force that causes DNA to take a helical shape?

   A. Hydrogen bonds between base pairs

   B. Aromatic base stacking

   C. Steric interactions

   D. Ionic charges on the phosphate backbone

7. **What is the nearest-neighbor approximation?**
   A. Approximating the DNA sequence of one organism with the organisms that live nearby.
   B. Assuming that the DNA sequence of one gene can only affect the genes that are immediately next to it.
   C. Ignoring interactions between nucleotides that are not immediately next to one another in the DNA sequence.
   D. Approximating the DNA sequence by looking at which nucleotides are next to one another.

8. **Which of the following statements is false?**
   S1. Superhelical DNA requires a force to hold the superhelix in place.
   S2. Circular DNA is always superhelical.
   S3. A superhelix is a quaternary structure.
   S4. Energy stored in the superhelix can be used to unwind the DNA.
   A. S1 and S2
   B. S1 and S3
   C. S1 and S4
   D. S2 and S3
   E. S2 and S4

9. **What is a topological invariant?**
   A. A property that does not change under smooth deformations of shape
   B. The writhe
   C. The linking number
   D. A and B
   E. B and C
   F. C and A

10. **A 5000-base-pair circular DNA has a linking number of 500, and its most stable amount of twist is 10 base pairs per turn. What is the writhe?**
    A. 10
    B. −10
    C. 0
    D. −0.1

chapter **11**

Membrane Biophysics

In Chaps. 7 and 8 we learned some basics of lipids and cell membranes. In this chapter we explore membrane biophysics. Biological membranes do more than just providing a barrier between the cell and the outside world, and they are made up of more than just lipids.

CHAPTER OBJECTIVES

In this chapter, you will

- Review the basic structure of biological membranes.
- Learn the principles that govern self-assembly in lipid bilayers.
- Learn about the fluid mosaic model.
- Study phase transitions in lipids and the forces that affect them.
- Gain an understanding of the energetics of membrane permeability.
- Learn about active and passive transport across membranes.

Membrane Functions

Membranes play important biological functions. They provide a barrier between the cell and the outside world, and regulate what can enter the cell and what leaves it. Many organelles within the cell are themselves membrane bound, providing an isolated biophysical environment even within the cell. In eukaryotes a membrane surrounds the nucleus, keeping the activities of DNA (transcription and replication) separate from the rest of the cell. Membranes also provide surfaces on which biochemical reactions take place. One such example is the *endoplasmic reticulum*, a network of membranes and tubules on which synthesis of proteins and lipids occurs. Membranes also provide anchorage points for the cytoskeleton which gives shape and rigidity to cells and plays a role in intracellular transport of biomolecules.

Membrane Structure

The main structural components of biological membranes are amphipathic lipids (see Chap. 7). Although biological membranes also contain carbohydrates and proteins (some membranes contain as much as 50% protein), the primary character of biological membranes is derived from its amphipathic lipids. The most common lipids found in biological membranes are two-chain phospholipids. These are molecules with a phosphate head group, attached to a two-chain fatty acid. The amphipathic character comes from the combination of the hydrophobic, hydrocarbon tails, along with the hydrophilic phosphate head group. In biological membranes, the phospholipids arrange themselves into a bilayer, in which the hydrocarbon tails face each other and are isolated from the surrounding environment by the phosphate head groups. See Fig. 11-1.

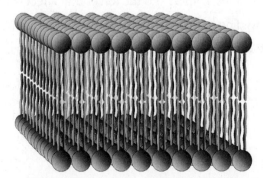

FIGURE 11-1 • The basic structure of a biological membrane is a phospholipid bilayer with hydrophobic tails in the center of the bilayer and hydrophilic heads at the surfaces. Biological membranes also typically contain proteins, nonpolymeric lipids such as cholesterol, and carbohydrates typically in the form of glycolipids (a carbohydrate attached to a lipid).

Phospholipid Behavior and Self-Assembly

Phospholipids exhibit a physical behavior that makes them ideal for the formation of membranes. Biophysicists call this behavior *self-assembly*. *Self-assembly* means that the molecules will aggregate together to form various structures without need for energy input, catalysts, or other helper molecules. Let's first look at self-assembly for single-chain phospholipids and then for double-chain phospholipids.

Single-Chain Phospholipids and Micelle Formation

If we slowly add a single-chain phospholipid to an aqueous solution, at first the lipid molecules are dispersed among the water molecules. Later, as the concentration of lipids is increased, a point is reached where the lipid molecules merge together forming aggregates called *micelles*. A *micelle* is simply a lipid ball with the hydrocarbon chains pointed in toward the center, as illustrated in Fig. 11-2. The concentration, at which the lipids self-assemble into micelles, is called the *critical micelle concentration* (CMC). The aggregation process is highly cooperative. Once the critical micelle concentration is reached, any further lipid added to the solution either associates with existing micelles or forms new micelles. This causes the concentration of micelles to increase, while the concentration of free lipid molecules remains relatively constant, as shown in Fig. 11-3.

There are a number of forces that contribute to self-assembly in micelle formation. As the hydrocarbon tails approach one another, dispersion forces (see Chap. 6) provide a strong attractive force pulling them together. Close association of the tails is also favored by the hydrophobic effect. As the tails are

FIGURE 11-2 • Single-chain phospholipids self-assemble into micelle structures.

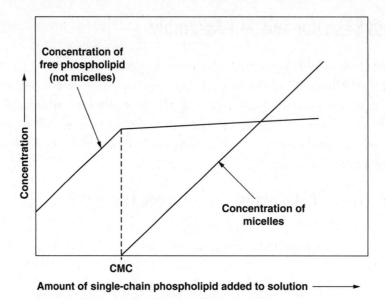

FIGURE 11-3 · As single-chain phospholipid is added to an aqueous solution, the concentration of free phospholipid increases, until the critical micelle concentration is reached. Above the CMC, micelle formation is highly cooperative, and additional lipid added to the solution increases the micelle concentration.

drawn together, water molecules along the length of the hydrophobic tails are disrupted (disorganized) and excluded (pushed away). Recall from Chap. 6 that water along the edge of a hydrophobic molecule has its motion restricted; this in turn limits its ability to hydrogen-bond with other water molecules. As the hydrocarbon tails come together, the water along their surface is pushed aside. This increases the entropy of the water by allowing it to rotate freely. The free rotation of the water molecules also increases the possibilities for hydrogen bond formation with other water molecules in solution. The increased entropy and the formation of water-water hydrogen bonds both contribute favorably to the Gibbs energy of micelle formation.

In addition to these forces there is a mixed effect from the negatively charged phosphate head groups. On the one hand, the charge on the head groups creates a repulsive force between them. This is, in part, what gives the micelle its spherical shape. As the hydrocarbon tails come together by virtue of the dispersion forces and the hydrophobic effect, we need to somehow reduce the repulsive force of the head groups. To do this, nature puts the head groups on the outside of the micelle and shapes the micelle like a sphere. This configuration maximizes the distance between the head groups; this reduces their repulsive interaction. It also allows space between the head groups for water molecules and counterions, both of which act to reduce the repulsive force. The water molecules do so by aligning their dipole moment opposite the electric field (as was explained in Chap. 6). If cations such as calcium, magnesium, or sodium are present, they associate as counterions with the negative charge on the phosphate. This effectively neutralizes the negative charge. Neutralizing the negative charges eliminates the need for energy (enthalpy) to overcome the repulsive force, so the Gibbs energy of micelle formation becomes even more negative, making it easier to form micelles. The critical micelle concentration is the concentration of lipid necessary for all of these forces combined to produce a negative Gibbs energy of micelle formation. At that point formation of micelles is highly cooperative, proceeding in an almost all-or-none fashion. The fact that counterions act to eliminate the repulsive force between head groups means that fewer lipid molecules are required to provide the energy (from dispersion forces and the hydrophobic effect) needed to achieve micelle formation. Fewer lipid molecules required means the critical micelle concentration is lower. This explains why, as the concentration of positive ions increases, the CMC decreases.

Two-Chain Phospholipids and Liposome Formation

Although micelles can be formed from two-chain phospholipids, this is usually not the preferred configuration. Compare Fig. 11-4 with Fig. 11-2. In the case of two-chain phospholipids, the hydrocarbon chains from adjacent lipid molecules are not able to pack as closely together as they can when all of the lipids have only a single hydrocarbon chain. The single-chain phospholipids get closer together both in terms of aligning side by side, as well as in terms of reaching deep into the center of the micelle. The extra width of the two-chain phospholipids causes steric interactions that increase the distance between hydrocarbon chains of adjacent molecules. This reduces the strength of dispersion forces. The extra width also blocks the hydrocarbon chains from reaching as deep into the micelle. This creates a void in the center of the micelle and increases the size of micelle, compared to a micelle made from the same number of single-chain lipid molecules. Both the void and the increased size (per number of lipid molecules) increase the energy cost of micelle formation. The void does so either because it will contain water molecules, thus putting them right next to the ends of the hydrophobic tails, or because somehow the water must be excluded from the center of the micelle without filling that space with something else (e.g., hydrocarbon, as with the single-chain lipids). Both of these are energetically unfavorable. The increased size of the micelle decreases its curvature.

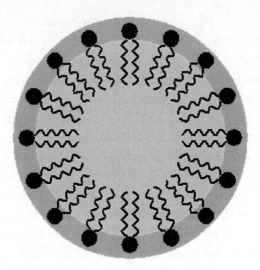

FIGURE 11-4 · A micelle made of two-chain phospholipids. In the micelle configuration, the lipid molecules are not able to pack as closely together as is the case with single-chain phospholipids. The lipid chains are also not able to reach as deep into the center of the micelle. This reduces the amount of contact between the hydrocarbon chains; this reduces the favorable energy from dispersion forces. These factors make micelle formation more difficult to achieve with two-chain phospholipids. Instead two-chain phospholipids prefer a bilayer configuration. (*Courtesy of Wikimedia Commons.*)

Given the same number of lipid molecules, a decreased curvature puts the repulsive phosphate head groups closer together. This is also unfavorable. So altogether it's a lot more expensive to make two-chain lipid micelles than it is to make single-chain lipid micelles.

Nature solves this problem by bringing a second layer of lipids opposite the ends of the hydrocarbon chains. It's as if the sides of the micelle split open, allowing the ends of the phospholipid chains across from each other to collapse into the void of the micelle and line up to form a *lipid bilayer*. This process is also highly cooperative, and it continues aggregating phospholipids in this way until a *liposome* has formed. A *liposome* is a lipid bilayer sphere with a void in the middle, as illustrated in Fig. 11-5. In this case, however, the void in the middle is favorable. It can be filled with water molecules and counterions which stabilize the liposome (contribute to a favorable Gibbs energy change) by interacting with the phosphate head groups. The interaction between water,

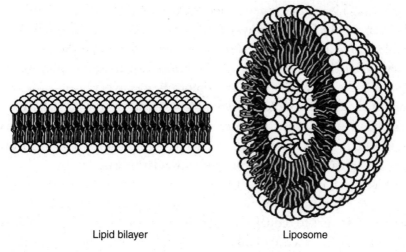

Lipid bilayer   Liposome

FIGURE 11-5 • Liposome formation. Two-chain phospholipids prefer liposomes to micelles because the energy cost is less.

counterions, and the negatively charged phosphates is the same as was described for micelles earlier. But in the case of a liposome, there are two surfaces of phosphate head groups on which this favorable interaction can take place.

Lipid Bilayer Energetics and Permeability

The lipid bilayer is *semipermeable* meaning that some molecules can pass through the bilayer while others can't. The ease with which any molecule can pass through the lipid bilayer depends on the Gibbs energy change of getting the molecule past the charged phosphate head groups and into the hydrophobic interior of the bilayer. After that we have the Gibbs energy change for removing the molecule from the hydrophobic interior and getting past the charged phosphate head groups on the other side of the bilayer. Both of these Gibbs energy changes depend largely on the strength of the forces that hold the bilayer together. They also depend on the interaction of these forces with the particular molecule attempting to pass through the bilayer. We'll look at the factors that affect the strength of these forces in a minute, but first let's take some general examples of molecules and how they interact with the bilayer.

Charged and polar molecules experience a significant increase in Gibbs energy just to penetrate the hydrophobic interior of the bilayer. As a result, they are typically not able to pass through the bilayer. (If a charged or polar molecule somehow has a very large kinetic energy, then it may be able to penetrate into the bilayer, at which point moving out of the bilayer to either side is highly favorable.) Large molecules are also unable to pass through the bilayer, unless they have enough energy to disrupt the forces that hold the bilayer together. Being large, they have to push aside more lipid molecules in order to pass through than a smaller molecule would have to do. So, they also have to have enough energy to disrupt more of the forces holding the lipid molecules together. On the other hand, small, uncharged, nonpolar (or only slightly polar) molecules can pass through the bilayer with relative ease.

The forces that hold the bilayer together are dispersion forces, the hydrophobic effect, and, if cations are present, an additional stabilization by the cations acting as counterions to reduce the repulsive force between the phosphate head groups. Obviously since cations stabilize the bilayer, their presence has a strengthening effect on the bilayer. This makes it more difficult for large molecules to enter, since large molecules have to somehow push the lipids aside. Pushing the lipids aside requires disrupting the dispersion forces that attract the hydrocarbon tails to one another. This takes energy, and the larger the molecule, the more lipid molecules it has to push aside (and the more energy required). On the other hand, at lower concentrations of cations, the repulsive force of the head groups may assist the large molecules getting through, by making it easier to push the lipids aside.

The strength of the dispersion forces depends largely on the amount of close contact between the hydrocarbon chains. This in turn depends on how close together the chains can pack and how long the chains are. Longer chains have more contact over which the dispersion forces can attract adjacent molecules. So, longer chains make stronger dispersion forces and more stable bilayers. (Longer chains also mean a thicker bilayer and thus a greater distance over which a molecule passing through the bilayer has to travel.) Another significant factor is whether the hydrocarbon chains are saturated, unsaturated, or polyunsaturated. Unsaturated chains (recall from Chap. 7) have one or more double bonds between some of the carbon atoms. Double bonds restrict free rotation, so each chain will be stiff at the location of the double bond, with a particular angle or kink in the chain at that location. The end result, due to these kinks, is that unsaturated and polyunsaturated chains prevent the lipid molecules from packing as closely together as saturated lipids can. This is illustrated in Fig. 11-6. Dispersion forces are proportional to $1/r^6$; that is, they decrease in proportion to the sixth power of the distance. Less tightly packed chains mean that on average the chains are further apart from each other; this leads to weaker dispersion forces.

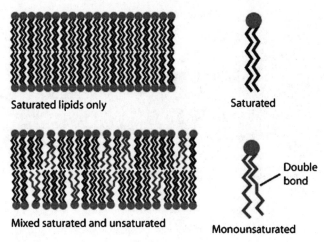

FIGURE 11-6 • Lipid bilayers that contain only saturated lipids are able to pack the lipids much closer together than bilayers that contain unsaturated lipids or a mixture of saturated and unsaturated lipids.

Still Struggling

The main forces that hold the bilayer together are the hydrophobic effect and dispersion forces. The hydrophobic effect is driven by entropy: water has more freedom of movement when it stays away from hydrophobic portions of the lipid molecules. Bilayer formation is favorable because it keeps the hydrophobic lipid tails away from the water, and the polar lipid head groups facing the water. Dispersion forces are due to synchronous fluctuations in the electron density of the atoms within the hydrophobic tails. The electron densities fluctuate rapidly and in synch with each other. At any given moment, in a place where one tail has a slightly positive charge an adjacent tail has a slightly negative charge. This causes the hydrophobic tails to attract one another giving the membrane strength and stability.

Fluid Mosaic Model

Our current working model of biological membranes is the *fluid mosaic model*, which was first proposed by Singer and Nicolson in 1972. The model states that biological membranes are composed primarily of a phospholipid bilayer. Other

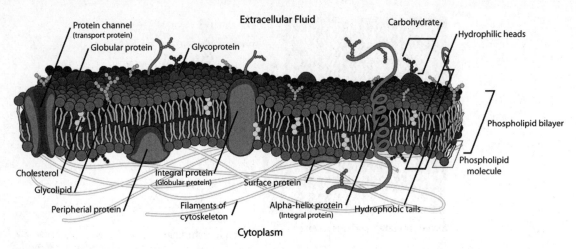

FIGURE 11-7 · Fluid mosaic model of biological Membranes. The phospholipids in the bilayer form a mosaic with other molecules making up the membrane. These molecules and the phospholipids are free to move about in two dimensions, that is, in parallel with the bilayer.

molecules that are part of the membrane structure, such as proteins, glycoproteins, cholesterol, and carbohydrates, form a mosaic with the lipid molecules. See Fig. 11-7. This is in contrast to earlier biomembrane models in which the lipids were thought to form a monolayer or in which proteins were also thought to form one or more layers within the membrane. In the fluid mosaic model, the lipids and other molecules are free to move about within the two-dimensional mosaic.

Phase Transitions in Phospholipid Bilayers

Phospholipids in bilayers and micelles can exist in a solid or fluid phase. We will focus our discussion on bilayers, since that is what we find in cell membranes, but the basic concepts apply also to micelles. The solid phase is called the *gel phase*. In the *gel phase* the bilayer is still somewhat flexible and permeable, but lipid molecules (and other molecules within the mosaic) are fixed to a particular position within the two-dimensional mosaic. In simple terms, they don't move around within the membrane.

The fluid phase is called the *liquid crystalline phase*. In the *liquid crystalline phase* phospholipids next to each other are able to change positions, sometimes as often as millions of times per second. In this way, any given lipid molecule is able to gradually diffuse and move about in two dimensions within the plane of the bilayer. However, in both the gel and the liquid crystalline phases, lipid molecules generally do not move from one layer to the other, since this would involve a large, positive Gibbs energy change. Proteins and other molecules in the mosaic also move about within the plane of the bilayer.

The Melting Transition

Phase transitions in lipid bilayers can be temperature induced. Therefore (as we saw with DNA unwinding) the transition is sometimes referred to as *melting*. The melting transition in phospholipid bilayers is highly cooperative, meaning that it tends to happen in an all-or-none, two-state manner. Figure 11-8 shows a typical melting curve for a sample of liposomes. The melting temperature T_m is the halfway point of the transition. Below the melting temperature the lipids are in the gel state; above the melting temperature they are in the liquid crystalline state. The transition results from the disruption of the dispersion forces among the hydrocarbon tails. The dispersion forces otherwise hold adjacent phospholipids together. The lipid bilayer itself remains intact due to hydrophobic forces, which can be further strengthened by counterions along the phosphate surfaces of the bilayer.

Biological Significance

Biological membranes are most commonly found to be in the fluid (liquid crystalline) state; this allows the free movement of molecules within the two-dimensional mosaic. However, the melting temperature is usually close to

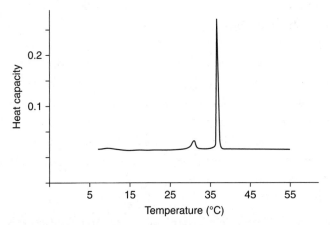

FIGURE 11-8 • Typical heat capacity melting profile for a highly cooperative gel to liquid crystal phase transition in a sample of liposomes (phospholipid bilayer vesicles).

physiological temperatures. This means that the cell can somewhat regulate the extent of gel versus liquid crystalline state of the membrane. In the liquid crystalline state, membranes are more permeable; this makes it easier for various molecules to pass into or out of the cell. Cells can decrease their overall permeability and increase the rigidity of their boundary by having some of the cell membrane in the gel state. However, if the *entire* cell membrane is in the gel state for a significant period of time, this typically results in the cell "freezing" to death. The ability of organisms to regulate the fluidity of their own membranes is called *homeoviscous adaptation*.

In the fluid state, the mobility of molecules within the mosaic plane enhances a cell's ability to move membrane proteins and receptors around to where they are needed. It also gives the cell the ability to reseal small holes in the bilayer, by shifting molecules around to fill in the hole. This important feature is lacking in gel phase bilayers. In some cases it may be necessary to keep some molecules on one particular side of the cell, for example, cilia (hairlike filaments) that extend from the surface of cells in the lung. These cilia could not do their job of sweeping dust out of the lungs if they were allowed to freely diffuse to the opposite side of the cell. On the other hand, keeping the cell membrane in the gel state would reduce permeability too much and prevent important molecules (e.g., oxygen) from getting through the cell membrane. The cell solves this problem by maintaining the bilayer in a liquid crystalline state, while anchoring some molecules in place with the cytoskeleton.

Factors Affecting Membrane Fluidity

The same factors that affect membrane permeability also affect membrane fluidity. Longer hydrocarbon tails increase dispersion forces making it more difficult to melt the bilayer. Experimentally we observe a higher melting temperature for bilayers made from lipids with longer hydrocarbon tails. Conversely, unsaturated lipids and lipids with shorter tails have lower melting temperatures due to weaker dispersion forces. In general, anything that weakens dispersion forces will reduce the melting temperature, increase fluidity, and increase permeability. Anything that strengthens dispersion forces will increase the melting temperature, decrease fluidity, and decrease permeability.

Cholesterol

Cholesterol is a significant component of many biological membranes, and it has a mixed effect on membrane fluidity depending on whether the bilayer is in the gel or liquid crystalline phase. Cholesterol tends to weaken dispersion forces by positioning itself between the phospholipid molecules. It also disrupts the orderly arrangement of the phospholipids in the gel phase, increasing entropy. Both of these effects lower melting temperature and favor the fluid phase of the bilayer.

However, in the fluid phase, the flat surface of cholesterol's interconnected rings (see Fig. 7-13) is significantly wider than the two chains of a single phospholipid molecule. And the interconnected rings are rigid. So each cholesterol molecule acts as an unbending barrier that limits some of the motion of the phospholipids. So, while the cholesterol prevents the formation attractive dispersion forces, the steric restriction of phospholipid movement decreases membrane fluidity and increases membrane rigidity. Many organisms are able to regulate membrane fluidity and rigidity by regulating the amount of cholesterol incorporated into their membranes. Organisms that live primarily in cold environments are typically found to have less cholesterol in their membranes. This prevents the organism from freezing to death.

The cholesterol molecule is almost entirely hydrophobic. Only the hydroxyl group at one end of the cholesterol molecule is hydrophilic. The hydroxyl group hydrogen bonds itself to the base of the phosphate head groups of the bilayer. Since the hydroxyl group is smaller than the phosphate head groups, the cholesterol is entirely submerged in the bilayer. The vast majority of the cholesterol molecule is comfortably inside the hydrophobic portion of the bilayer.

Membrane Growth

As cells grow, their membranes need to grow also. The cell somehow needs to add lipid molecules to the bilayer. Lipids are synthesized in the cell interior through a series of biochemical reactions mediated by enzymes. The resulting lipids aggregate to form liposomes. The lipids are then added to the membrane by fusing the liposomes with the existing membrane. See Fig. 11-9.

As their membranes grow, cells need to incorporate additional protein and other molecules that are part of the membrane structure. This can be done in a variety of ways. Some relatively hydrophobic proteins and glycolipids are able to simply penetrate the bilayer due to a favorable Gibbs energy change. Once inside the bilayer the hydrophobic effect helps to keep them in place. Sometimes proteins are inserted into the cell membrane while still in the form of a polypeptide chain, before the protein folds into its native state. The fact that protein synthesis typically takes place just below the surface of the cell membrane facilitates this process. Once the protein is inside the bilayer, the Gibbs

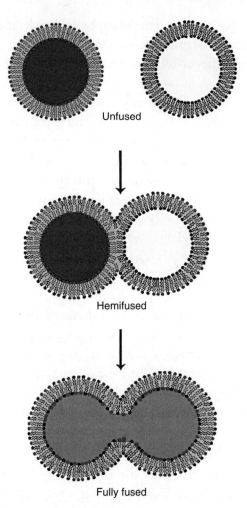

Unfused

Hemifused

Fully fused

FIGURE 11-9 · Fusion of two liposomes. The process is the same for a liposome fusing with an existing larger membrane.

energy change of protein folding becomes favorable. At the same time, the folding of the protein typically further stabilizes its position in the bilayer making it unfavorable for the protein to exit the bilayer.

Membrane Permeability and Transport

Let's revisit the idea of bilayer permeability and the need to transport molecules across membranes. As we noted earlier in this chapter, the lipid bilayer is *semipermeable*, meaning that some molecules can pass through the bilayer while others can't. Sometimes biological membranes are described as *selectively permeable*, meaning that the cell somehow *chooses* what to let in, what to let out, what to keep in, and what to keep out. The cell does this with a combination of *active* and *passive* transport. *Active transport* means that the cell expends energy to move a molecule or ion across the membrane that would otherwise not favorably move across the membrane. *Passive transport* means that the molecule or ion passes through the membrane by virtue of a favorable Gibbs energy change without the need for the cell to expend any energy.

Passive Transport

In the case of passive transport there are two important points to keep in mind. The first is that sometimes passive transport is the result of some earlier active transport. In other words, the cell regularly expends energy moving molecules against an otherwise unfavorable Gibbs energy change. Later, when the needs of the cell require it, the cell allows the molecules to flow in the direction of a favorable Gibbs energy change via passive transport. This type of passive transport may be for the purpose of simply adjusting the concentration of an ion or molecule. But that is usually not the case. Rather this type of passive transport is more commonly used as the driver of a *secondary active transport*. In secondary active transport, the energy released from a passive transport is used to drive an active transport.

The second point to know regarding passive transport is that it is not always a simple diffusion or movement through the lipid bilayer (combined with a favorable Gibbs energy change). Instead, passive transport can occur as the result of a tunnel, or channel, created through the membrane by the presence of *transport protein*. An example is shown in Fig. 11-10. Transport proteins can be involved in either active or passive transport. Transport proteins involved in passive transport contain a combination of hydrophobic and hydrophilic amino acid residues. The sequence of resides and the forces involved in protein folding cause the protein to fold in a way that creates a tunnel or channel through the center of the folded protein. The hydrophobic residues are in the outside of the protein exposed to the hydrophobic portion of the bilayer. The hydrophilic residues are on the inside of the protein, away from the hydrophobic lipid chains. These hydrophilic residues line the walls of the channel through the center of the protein. There are also usually some hydrophilic residues near the openings of the channel at each end of the transport protein. These may form hydrogen bonds with the phosphate head groups and help to keep the protein oriented so that the transport channel remains perpendicular to the plane of the bilayer (see Fig. 11-10).

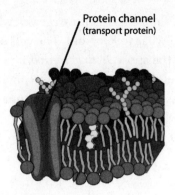

Protein channel
(transport protein)

FIGURE 11-10 · A channel is a type of transport protein that folds in a way so as to form a hydrophilic channel or tunnel through the phospholipid bilayer.

Gated Ion Channels

Earlier in this chapter we pointed out that lipid bilayers are very much *not* permeable to ions. In almost all cases, transport of ions across the membrane requires the existence of an *ion channel*, which is a transport protein as we described above with a hollow channel lined with hydrophilic residues. (One exception is vesicle transport discussed later in this chapter.) In some cases the native state of the membrane transport protein involves two or more native conformations. In at least one of these conformations the transport channel is blocked, preventing the movement of ions into the channel through either steric or hydrophobic forces. Such a channel is said to be *gated* because the conformation of the protein acts as a gate that can open and close. Depending on the particular transport protein and its specific purpose, various factors can act to trigger the conformational change to open or close the gate. For example, binding of a protein or other molecule to a specific receptor on the surface of the membrane may trigger the gate to open or close. Or the presence of a certain voltage difference across the membrane may trigger the conformational change that opens or closes the ion channel. This latter situation is referred to as a *voltage gated ion channel*. Voltage gated ion channels are common in nerve cells (neurons) and other excitable tissue (muscle cells). Typically there is a very specific voltage at which the gated ion channels of a cell open in a highly cooperative manner, so that once one channel is open, the other channels along the length of the neuron open rapidly in succession, creating an impulse that travels down the length of the neuron.

Active Transport

Active transport always involves the use of energy to drive the transport of a molecule or ion across the membrane. There are two reasons why energy may be needed to drive the transport of a molecule or ion across the membrane. The first reason is that there may be an otherwise unfavorable Gibbs energy change just to get the molecule into the hydrophobic interior of the bilayer. This is the case with ions and other charged or large and highly polar molecules. The overall Gibbs energy change to get the ion from one side of the membrane to the other may be small or even negative (favorable), but there is an energy hump, or barrier, to get over in order to get the ion into the bilayer. This energy barrier is illustrated in Fig. 11-11. In most cases, if the overall Gibbs energy change is

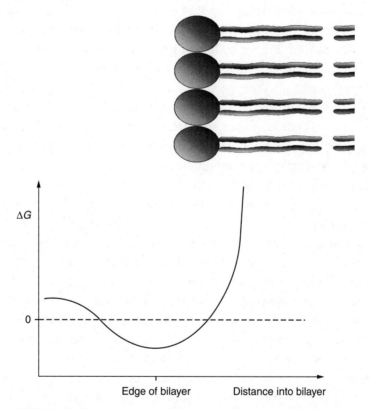

FIGURE 11-11 • Gibbs energy change for passing a positive ion into a lipid bilayer, as a function of distance into the bilayer. The Gibbs energy change is initially negative as the ion approaches the negatively charged phosphate head groups. If the concentration of the ion is smaller on the other side of the bilayer, then the overall Gibbs energy change will be negative (favorable). But as the positive ion passes the lipid head groups and into the hydrophobic interior of the bilayer the Gibbs energy rises sharply.

negative, then the cell can solve the energy barrier problem using a passive ion channel, as described previously. However, in some cases, especially those which involve relatively large molecules, the cell may make use of active transport to move the molecule into the bilayer and across.

The second and more common reason active transport is needed is because the cell requires a higher concentration of some substance on one side of the membrane. Purely passive transport would tend to equalize the concentrations, until the concentration is the same on both sides of the membrane. Gibbs energy changes are typically negative for moving a molecule from a region of high concentration to a region of low concentration (primarily for entropic reasons). So the Gibbs energy change favors equalizing the concentrations. This is why moving a molecule from a region of low concentration into a region of high concentration requires the input of energy to go against the concentration gradient. An example of this is the uptake of glucose from digested food. The glucose has to move from the hollow of the small intestine (where its concentration is low) into the cells that line the walls of the small intestine. The cells lining the wall of the intestine need to concentrate the glucose on their insides and then pass the glucose into the bloodstream. Without active transport, as soon as the concentration of glucose in the cells reached the level in the hollow of the intestine, no more glucose would flow into the cells. A significant amount of glucose in the digested food would be lost.

There are two types of active transport. Both are mediated by transport proteins that are part of the cell membrane. *Primary active transport* takes energy from cleaving a high energy phosphate bond and uses that energy directly to transport a molecule or ion across the membrane. *Secondary active transport* utilizes energy from concentration gradient to drive the active transport; that is, the favorable Gibbs energy of some molecule moving from high concentration to low concentration (a passive transport) is coupled with actively pushing some other molecule across the membrane. The term *secondary active transport* comes from the fact that the cell expended energy, via primary active transport, to create the concentration gradient in the first place.

Active transport that involves transporting more than one type of molecule or ion is called *cotransport*. Transport proteins that mediate the simultaneous transport (cotransport) of more than one type of molecule or ion are called *cotransporters*. Cotransporters that move the different molecules in the *same* direction across the membrane are called *symporters*. Cotransport proteins that move different molecules in *opposite* directions across the membrane are called *antiporters*. Most examples of cotransport that have been studied involve only two different molecules or ions. However, some cases of three or more have been studied. In cases where three or more molecules are moved by a cotransporter, if at last one substrate moves in the opposite direction to the others then the cotransporter is termed an *antiporter*. Only if all substrates move in the same direction do we use the term *symport*.

Still Struggling

Most books use the terms *symport* and *antiport* as defined here. However, some books use *cotransport* to mean "symport," and *countertransport* to mean "antiport."

Primary Active Transport

In *primary active transport*, the transport protein couples a reaction hydrolyzing ATP with moving the ion or molecule across the membrane. Recall from Chap. 7 that hydrolysis of ATP (adenosine triphosphate) involves cleaving one of the phosphate groups from ATP to form ADP (adenosine diphosphate). When the pyrophosphate bond is cleaved, a large amount of energy is released. This energy can be used by any process that is coupled to the reaction cleaving the phosphate bond. The coupling of a process to a reaction that involves hydrolysis of ATP is typically done by an enzyme classified as an *ATPase*, and the coupling is usually carried out by some conformational change in the enzyme. In the case of primary active transport, it is the transport protein itself that couples the hydrolysis with transporting a molecule or ion across the membrane.

The mechanism of active transport involves a conformational change in the transport protein. The conformational change is needed to push the molecule (or ion) through the hydrophobic bilayer or to move the molecule (or ion) against a concentration gradient. In primary active transport, without the hydrolysis of ATP the conformational change is otherwise not favorable. The enthalpy from the hydrolysis of ATP tips the Gibbs energy change in favor the conformational change that pushes the molecule across the membrane.

The Sodium-Potassium Pump

Figure 11-12 illustrates an example of primary active transport. This example is called the *sodium-potassium pump*. The term *pump* is used whenever a transport

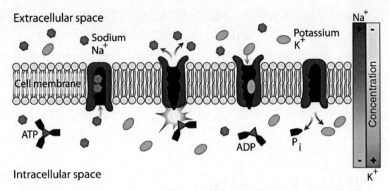

FIGURE 11-12 · Example of primary active transport: sodium-potassium pump.

protein actively moves molecules from a region of low concentration to a region of high concentration. The sodium-potassium pump establishes and maintains a concentration gradient such that sodium ions are concentrated on the outside of the cell, and potassium ions are concentrated on the inside of the cell.

The sodium-potassium pump is an example of a cotransporter. The transport protein is called *sodium-potassium ATPase* when referring to its enzymatic activity (otherwise we can also just call it the *sodium-potassium pump*). Sodium-potassium ATPase is an antiporter because it pumps sodium and potassium ions in opposite directions across the membrane. The mechanism of the sodium-potassium pump has been studied extensively and is a good example of a biophysical process that involves a complex combination of conformational changes, binding of multiple ligands, and enzymatic activity. Such biophysical processes are not uncommon.

Table 11-1 outlines the steps in the sodium-potassium pump mechanism. The protein has several binding sites for ligands. There are three sites that bind sodium ions, and two sites for binding potassium ions. There is also a site that binds ATP and catalyzes the hydrolysis of ATP. When the protein hydrolyzes ATP, it becomes *phosphorylated*. That is, the cleaved phosphate from the ATP becomes covalently attached to one of the amino acids in the protein. This phosphorylation is temporary and reversible. *Phosphorylation* of proteins is a common part of many biochemical processes. It can cause an otherwise hydrophobic amino acid to become charged or highly polar. This in turn will significantly alter the forces and various Gibbs energy interactions that determine the protein's secondary and tertiary structure. You can just imagine what can happen when a conformation that keeps the hydrophobic residues together and isolated from the aqueous environment, suddenly finds that one of the

TABLE 11-1 Mechanism of sodium-potassium ATPase (see also Fig. 11-13).

Step	Description
1	The pump (sodium–potassium ATPase) binds ATP and three intracellular Na⁺ ions.
2	Hydrolysis of ATP and phosphorylation of the pump. Conformational change exposes Na⁺ ions to the outside, reduces the pump's affinity for Na⁺, and increases the pump's affinity for K⁺. The three sodium ions are released outside the cell.
3	Two extracellular K⁺ ions are bound, triggering dephosphorylation.
4	Dephosphorylation triggers a reversion to the original conformation. Ion binding sites are exposed to the inside of the cell, the affinity for Na⁺ is strong, and the affinity for K⁺ is weak. Two K⁺ ions are released inside of the cell and the cycle repeats.

hydrophobic residues has become charged and very hydrophilic. The result is that phosphorylation can induce dramatic changes to a protein's conformation.

In its initial state (left side of Fig. 11-12) the protein's conformation is such that its three sodium binding sites and two potassium binding sites are exposed to the inside of the cell. In this conformation, the Gibbs energy change for binding ATP and for binding sodium ions is highly negative (favorable). At the same time, potassium binding is not energetically favorable. So the protein quickly binds ATP plus any free sodium ions that are nearby. Once three sodium ions are bound, the protein undergoes a slight conformational change that allows it to hydrolyze ATP. The ATP hydrolysis results in phosphorylation of the sodium-potassium pump. This in turn, as we mentioned, changes the Gibbs energy interactions that determine the secondary and tertiary structure of the pump. The phosphorylated form of the protein has a dramatically different conformation and a dramatically different affinity for binding sodium and potassium. In its new conformation (second illustration from the left in Fig. 11-12) the sodium and potassium binding sites are now exposed to the outside of the cell. But the Gibbs energy change for binding sodium ions is now positive (unfavorable). This means that releasing the sodium ions provides a negative (favorable) Gibbs energy change. So the three sodium ions are quickly released to the outside of the cell. All of this is the result of the changes in location of the various charges, polar groups, hydrogen bonds, and hydrophobic residues in the protein, due to the phosphorylation and accompanying conformational change.

In its new phosphorylated conformation, the protein's affinity for potassium is significantly increased. In other words, in this new conformation, binding of potassium is accompanied by a negative Gibbs energy change. This is illustrated in the third depiction of the sodium-potassium pump (third from the left in Fig. 11-12). Binding of potassium is coupled to a highly favorable dephosphorylation of the protein, so that binding potassium ions triggers the dephosphorylation. The phosphate separates from the protein and the protein reverts to its initial conformation. Since the initial conformation has a low affinity for potassium ions, the potassium ions separate from the pump and enter the interior of the cell, and the cycle repeats.

Secondary Active Transport

Secondary active transport utilizes energy from a concentration gradient to drive an active transport. If the concentration gradient is steep enough, then the Gibbs energy change for moving a molecule from high concentration to low concentration will be very negative. This very negative Gibbs energy change can be coupled with a transport that would otherwise have a somewhat positive (unfavorable) Gibbs energy change. The total Gibbs energy change for the coupled reaction will be negative. In this way, a highly favorable passive transport can be coupled with an otherwise unfavorable transport, actively driving it forward.

The cell is expending energy, so the transport is active. But the energy comes from energy stored in the concentration gradient. The concentration gradient got this energy from a primary active transport that actively pumped ions (or molecules) from one side of the membrane to the other, thereby creating a large concentration difference. Since the concentration difference originated due to a primary active transport, an active transport driven by this concentration difference is termed *secondary active transport*.

Secondary active transporters, by definition, are always cotransporters, since it is specifically the passive transport of one type of molecule or ion that is driving the secondary active transport of another molecule or ion. The cotransport by a secondary active transporter can be symport (in the same direction) or antiport (in the opposite direction).

Many secondary active transporters are linked to the strong sodium gradient created by sodium-potassium ATPase. Examples include the symport of glucose and neutral amino acids. Sodium-linked symport proteins that drive the transport of glucose and neutral amino acids are commonly found in intestinal cells and kidney cells. Although the Gibbs energy change is favorable for sodium ions to move into the cell, they still need to get through the hydrophobic cell membrane. In those cells that employ sodium-linked symport proteins, the sodium binds to the transporter protein as does the cotransported molecule (glucose or the neutral amino acid). Only when all of the cotransported molecules are bound can the sodium move with the concentration gradient to drive the conformational change needed to push the cotransported molecule inside. Once inside both are released, allowing the transport protein to return to its original state. In the case of sodium-linked symport of glucose, two sodium ions move across the cell membrane for every glucose molecule transported.

An example of a sodium-linked antiporter is the *sodium-calcium exchanger*. This membrane transport protein moves three Na^+ ions into the cell for every Ca^{2+} ion that it moves out of the cell. The *sodium-calcium exchanger* protein plays an important role in maintaining low calcium ion levels in cardiac (heart) cells, by pumping the calcium ions out of the cell. In cardiac and other muscle cells, increased levels of intracellular calcium cause muscle contraction. Keeping the calcium levels low allows the muscles to work properly without excessive contraction. However, in the case of a weakened heart muscle, drugs may be used to increase the strength of heart muscle contractions. Some such drugs work by partially inhibiting sodium-potassium ATPase in cardiac cells. This reduces the sodium gradient; this reduces the driving force for the sodium-calcium exchanger and so allows intracellular calcium concentrations to increase. The increased intracellular calcium increases the strength of cardiac muscle contraction.

Vesicular Transport

Another term for liposome is *lipid vesicle*. *Vesicular transport* involves liposomes that contain some molecules in their hollow center, fusing with, or budding off of, a membrane. In this way the contents of the liposome are transferred to the other side of the membrane without ever having to actually pass directly through the lipid bilayer. We saw fusion of liposome membranes in Fig. 11-9, when we were discussing membrane growth. Budding, or *extrusion*, occurs when a portion of a membrane is pushed out and eventually pinches off to form a new liposome. This process is illustrated in Fig. 11-13. Vesicular transport is not only a method of getting molecules across a membrane, but is also used extensively by cells to transport molecules around within the cell, from one membrane organelle to another, and even externally from cell to cell.

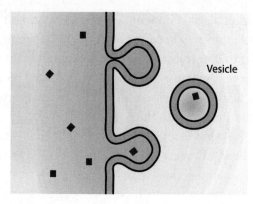

FIGURE 11-13 • Budding (extrusion) of a membrane to form a lipid vesicle.

Note that *budding* and *fusion* work together. Both move molecules from one side of a membrane to another. And both can move molecules either into or out of a cell. But budding traps the molecules inside a liposome as they cross the membrane, as we saw in Fig, 11-13. In this way budding is a preparation for transporting the molecules further, not just to the other side of the membrane. Fusion, however, sets the molecules free on the other side. See Fig. 11-14.

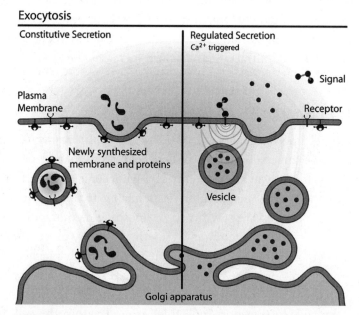

FIGURE 11-14 • Fusion of a lipid vesicle with a membrane releases the contents of the vesicle to the other side of the membrane.

Vesicular transport is a complicated form of transport with many variations, many aspects of which are only beginning to be understood in depth. A full treatment of vesicular transport is beyond our scope; however, we will highlight some of the major features here.

Common uses of vesicular transport include

- Secretion of enzymes, hormones, and other substances out of secretory cells
- Waste elimination
- Intake of nutrients
- Cellular digestion of nutrients
- Uptake of foreign bodies for destruction
- Viral entry into cells
- Transport of proteins from one organelle to another or from one location in the cell to another
- Delivery of drugs for treatment of disease

Vesicular transport *into* a cell is called *endocytosis, out* of a cell is called *exocytosis.* Endocytosis is further broken down into several categories depending on what is coming into the cell. For example, *phagocytosis* refers to the uptake of solid particles into the cell. *Pinocytosis* is the intake of liquids and dissolved substances. *Receptor-mediated endocytosis* means that receptors on the surface of the membrane bind very specific molecules (such as hormones or proteins) which are then carried through the membrane into the cell via budding. Receptor-mediated endocytosis is a good example of the cell membrane being selectively permeable.

Some of the stages of vesicular transport require the cell to expend energy, and so vesicular transport can be classified as an active transport. For example, budding requires the polymerization of microtubules (components of the cytoskeleton) in order to push or pull a section of membrane out into a bud. The polymerization process requires the input of energy. Vesicle fusion often requires assistance by ATP-dependent enzymes. In other words, the proteins that help begin the process of vesicle fusion couple the fusion reaction with ATP hydrolysis in order to provide the energy needed to make the process favorable. An example of this is exocytosis of neurotransmitters (chemicals that carry nerve signals from one nerve cell to another). After the vesicle containing

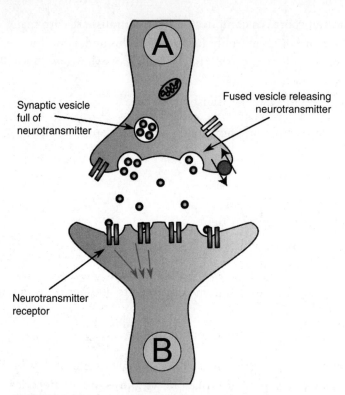

FIGURE 11-15 • Exocytosis of a neurotransmitter from cell A to cell B. The fusion of neurotransmitter vesicles is mediated by an ATP-dependent enzyme, and so makes this an active transport (one in which the cell expends energy).

neurotransmitter molecules binds to the neuron cell membrane, an ATP-dependent enzyme makes the necessary modifications to the lipid bilayer in order for membrane fusion to occur. Figure 11-15 illustrates the exocytosis of neurotransmitter molecules.

QUIZ

Refer to the text in this chapter if necessary. Answers are in the back of the book.

1. Most biological membranes are made from what?
   A. Single-chain phospholipids
   B. Two-chain phospholipids
   C. Three-chain phospholipids
   D. Four-chain phospholipids

2. Which statement is true?
   A. Lipid bilayers can self-assemble.
   B. Cholesterol blocks self-assembly.
   C. Micelles cannot self-assemble.
   D. Large hydrophobic proteins stimulate self-assembly.

3. What are the major forces that hold membranes together?
   A. Phospholipid stacking and induced dipoles
   B. Hydrogen bonds and hydrophobic interactions
   C. Dispersion forces and hydrogen bonds
   D. Hydrophobic interactions and dispersion forces

4. How can large molecules pass through a membrane?
   I. Vesicle fusion
   II. Endocytosis
   III. Active transport
   IV. Passive transport
   V. Optical tweezers
   A. I, II, and III
   B. II, III, and IV
   C. III, IV, and V
   D. II, III, and IV

5. Where does the energy come from for primary active transport?
   A. Ionic currents
   B. ATP
   C. Temperature gradient
   D. Concentration gradient

6. Where does the energy come from for secondary active transport?
   A. Ionic currents
   B. ATP
   C. Temperature gradient
   D. Concentration gradient

7. **Which statement is *most* true?**

   A. The sodium-potassium pump is an active transport membrane.
   B. The sodium-potassium pump is a protein.
   C. The sodium-potassium pump is an ion channel.
   D. The sodium-potassium pump is a passive transport protein.

8. **What role does cholesterol play in membranes?**

   A. It decreases membrane fluidity.
   B. It increases membrane rigidity.
   C. It creates steric restrictions on phospholipid movement within the fluid mosaic.
   D. All of the above
   E. None of the above

9. **Why does the presence of unsaturated lipids lower the melting temperature of liposomes?**

   A. The double bonds stiffen the hydrocarbon chains; this spreads the lipids apart and reduces dispersion forces.
   B. The double bonds are easier to melt because of a more favorable contribution to the Gibbs energy change.
   C. Liposomes don't melt.
   D. The double bonds reduce the cooperativity of the phase transition by synchronizing the charge fluctuations of the phospholipid head groups.

10. **What happens to the heat capacity when lipids undergo a phase transition from the gel state to the liquid crystal state?**

    A. Nothing.
    B. The heat capacity rises sharply because the heat capacity of the liquid crystal state is higher than that of the gel state.
    C. The apparent heat capacity rises during the transition, as energy is used to disrupt dispersion forces, and then falls when the transition is complete.
    D. The heat capacity makes a favorable contribution to the Gibbs energy change of bilayer formation.

chapter 12

Physiological and Anatomical Biophysics

We now turn our attention to physiological and anatomical biophysics; biophysics at the level of the entire organism (multicellular organisms) or organs and systems within an organism. We will only touch on a few aspects of this very broad area of biophysics in order to give you a feel for what is involved.

CHAPTER OBJECTIVES

In this chapter, you will

- Learn some basic concepts in physiological and anatomical biophysics.
- Determine the height of a jump given the weight of an animal and the force generated by its legs.
- Calculate the velocity of blood in the aorta.
- Come to understand how arteriosclerosis affects blood flow.
- Study some aspects of the aerodynamics of hummingbird flight.

The Scope of Physiological and Anatomical Biophysics

In this chapter we explore just a few aspects of physiological and anatomical biophysics. Physiological and anatomical biophysics is enormous, big enough to fill a volume by itself, or more. It includes basic mechanics—static forces, dynamic forces, and all types of motion—as applied to organisms and to parts of organisms. It includes fluid dynamics both as an analysis of fluids inside organisms (e.g., blood) and as an analysis of organisms that live in or spend time in fluids (e.g., birds in the air and fish in water).

Physiological and anatomical biophysics also includes acoustics (the physics of sound) and optics (the physics of light) as they relate to hearing and seeing, and the organs and mechanisms involved in hearing and seeing. Just to name a few more things, we can also include heat and energy and how an organism controls its use of energy and its temperature. And we can include the study of materials strengths and elasticity, as it relates to the various parts of an organism, for example, muscles and bones, or stems and leaves, and other structures that require strength and elasticity.

Jumping in the Air

How much energy does it take to jump in the air? How high can you jump? The basic principles of jumping apply whether the jumping organism is a person jumping for joy, a mountain lion leaping for its prey, or a dolphin jumping out of the water. The physics of jumping is easily broken into three steps. In step 1 an organism, with no initial upward velocity, accelerates upward. In step 2 the organism is in the air with an upward velocity and a downward acceleration due to gravity. The force of gravity slows the upward velocity, eventually reaching zero at the peak of the jump. In step 3 the organism falls back down, with an increasing downward velocity as a result of the acceleration due to gravity.

Let's look at the first two steps, the jump itself, in a little more detail. As a reminder (if you've studied basic physics) the motion of any object can be broken down into separate motions in each of three dimensions: up and down, forward and backward, left and right. It's much simpler to deal with motion in a single direction at a time (ignoring the others) and mathematically the results are absolutely correct.

Let us say that initially our organism has no upward or downward motion (although the organism may be running or swimming forward). The organism then exerts a downward force against a medium (e.g., the ground or water). The result is an equal and opposite upward force against the organism (Newton's third law of motion). This force creates an upward acceleration, increasing the organism's upward velocity for as long as the force is applied. At some point the organism leaves the medium (the ground or the water) and the upward force stops. This is the end of step 1.

The acceleration of step 1 has brought the organism to some maximum upward velocity. This maximum upward velocity is the initial velocity for step 2. In step 2 the organism is in the air with an initial upward velocity and a downward acceleration due to gravity. The force of gravity accelerates the organism downward, gradually reducing the upward velocity to zero. The organism continues to rise upward in the air until the upward velocity is zero. At this point the organism is at the peak of its jump. This is the end of step 2.

Let's take an example of a basketball player making a jump shot. Most organisms that jump from the ground will crouch down, bending their knees just prior to the jump in order to apply force with their leg muscles. As the force is applied, the legs straighten. The force can only be applied for as long as the legs are bent. Once the legs are straight, the organism lifts off the ground and step 2 begins. Two things affect the initial upward velocity of step 2, the size of the upward force and the length of time over which the force is applied. The length of time, in turn, depends somewhat on how far down the organism crouched. See Fig. 12-1.

FIGURE 12-1 · Left: A basketball player crouches down before a jump shot. Center: An upward acceleration is applied as the player straightens up, but only while the player is still on the ground. Right: Once the player is in the air, the upward force stops. Then the upward velocity begins to decrease as a result of the negative acceleration of gravity. The player reaches maximum height when the upward velocity reaches zero.

PROBLEM 12-1 _____

A 180-lb, 6-ft basketball player makes a jump shot. Just before jumping, the basketball player crouches down 8 in. If the player's leg muscles together apply an average force of 450 lb, how high will the basketball player jump?

SOLUTION _____

To begin with, let's convert to metric units, rounding off to keep the numbers relatively simple. 180 lb is 800 N. The fact that the player is 6 ft tall is irrelevant (at least to us, perhaps not to the coach). Eight inches is 0.2 m. And 450 lb of force is 2000 N.

In a real-life situation, the force applied would not be constant. However, to keep things simple we can use the average force and assume a constant force (equal to the average force) during the time the basketball player is accelerating up from the bent-knee position. Recall that force equals mass times acceleration ($F = ma$), so a constant force means a constant acceleration. And a constant acceleration means that the velocity is increasing linearly (in a straight line). This last point is helpful, because it means that wherever we have a constant acceleration, instead of having to figure out the increasing velocity at each point in time, we can simply take the average of the initial and final velocity, and apply that average velocity over the same period of time and get the same mathematical result.

We break the problem up into the same steps as previously. The first step is to determine maximum velocity at the end of step 1. Then using this as the starting velocity of step 2 and applying the downward acceleration of gravity, we calculate the distance traveled until the velocity is zero: This is the height of the jump.

Before we solve the problem, let's review and derive some basic formulas of motion. By definition, speed or velocity is the distance traveled divided by time. See Eq. (12-1). This will be used to relate the velocity of the jump to the height of the jump. (Throughout this section we will use speed and velocity interchangeably, Strictly speaking, velocity and acceleration are vectors, and speed is scalar. But for our purposes we are only dealing with motion in the up and down direction. Therefore to keep things simple, we will treat velocity and acceleration as scalars and use positive value for the up direction and negative value for the down direction.)

$$v = d/t \qquad\qquad (12\text{-}1)$$

The standard units are meters per second.

By definition, acceleration is the change in velocity divided by time.

$$a = (v_2 - v_1)/t \qquad\qquad (12\text{-}2)$$

Standard units are meters per second per second, or m/s^2.

When a force is applied to an object, the result is an acceleration that is equal to the force divided by the mass. This is Newton's second law and is usually written as force equals mass times acceleration.

$$F = ma \qquad\qquad (12\text{-}3)$$

The standard units of force are $kg\ m/s^2$, also called a newton.

In step 1 of our problem we don't know the time over which the force is applied and we don't know the resulting velocity (that's what we're trying to calculate), so we're going to have to play with the above equations to get them into forms that contain what we do know. We know the distance over which the force was applied (the 8-in, or 0.2-m, crouch), and we know the force and the mass which together can be used to calculate the acceleration.

Let's rearrange Eq. (12-2) to put it in terms of what we want to calculate in step 1.

$$v_2 = v_1 + at \qquad\qquad (12\text{-}4)$$

This says that if an object is initially traveling at velocity v_1 and we apply an acceleration a for t seconds, then the object will accelerate from velocity v_1 to velocity v_2. If the acceleration itself is changing, then we have to use integral calculus to calculate the average velocity during those t seconds. But if the acceleration is constant, then the velocity change is linear (i.e., a graph of velocity over time would be a straight line), so the average velocity is simply the average of the initial and final velocities.

$$v_{avg} = (v_1 + v_2)/2 \qquad\qquad (12\text{-}5)$$

Knowing the average velocity comes in handy because we don't know t (the amount of time the player is accelerating upward). Rearranging Eq. (12-1), we can express the time in terms of the distance and the average velocity (which we do know).

$$t = d/v_{avg} \qquad\qquad (12\text{-}6)$$

Substituting Eq. (12-5) into Eq. (12-6), we get

$$t = \frac{d}{v_{avg}} = \frac{d}{\frac{1}{2}(v_1 + v_2)} = \frac{2d}{(v_1 + v_2)} \tag{12-7}$$

Then substituting Eq. (12-7) into Eq. (12-4), we get

$$v_2 = v_1 + a\frac{2d}{v_1 + v_2} \tag{12-8}$$

Now multiply both sides of Eq. (12-8) by $(v_1 + v_2)$.

$$v_2(v_1 + v_2) = v_1(v_1 + v_2) + a\,2d$$

or

$$v_2^2 + v_1 v_2 = v_1^2 + v_1 v_2 + a\,2d$$

Then subtract $v_1 v_2$ from both sides to get Eq. (12-4) in terms of velocity, acceleration, and distance (instead of velocity acceleration and time).

$$v_2^2 = v_1^2 + 2ad \tag{12-9}$$

Now, in order to calculate the final velocity for step 1, all we need is Eqs. (12-3) and (12-9).

From Eq. (12-3) we know $a = F/m$. The vertical forces on the basketball player are the downward force of his weight (800 N) and the upward reactive force from the push of his leg muscles (2000 N). The net force is then 2000 N − 800 N = 1200 N. His mass is 180 lb/2.2046 lb/kg = 81.65 kg. The upward acceleration is then 1200 N/81.65 kg = 14.7 m/s².

Now that we know the acceleration, we can use Eq. (12-9) to calculate the final velocity for step 1. The initial vertical velocity at the beginning of step 1 is zero. The distance according to the problem is 0.2 m. Substituting these values into Eq. (12-9), we get

$$v_2 = \sqrt{0 + (2 \cdot 0.2 \cdot 14.7)} = 2.42 \text{ m/s} \tag{12-10}$$

Now we use the final velocity from step 1 for the initial velocity of step 2. In step 2 our basketball player is rising in the air, with an initial velocity of 2.42 m/s. The final velocity, at the peak of his jump, is zero. The only force on the player (ignoring air resistance and assuming no one bumps into him) is the force of gravity. The acceleration due to gravity is -9.807 m/s². Notice that the acceleration due to gravity is negative. That is because it is in the downward direction, and we have used the convention that upward distances, velocities, and acceleration are all positive. The distance traveled is the height we are looking for. So substituting these numbers into Eq. (12-9) and solving for d, we get

$$0 = (2.42)^2 + 2\,(-9.807)\,d \qquad\qquad (12\text{-}11)$$

Solving for d gives us

$$d = (5.86 \text{ m}^2/\text{s}^2)/(19.6 \text{ m/s}^2) = 0.3 \text{ m} = 11.7 \text{ in}$$

So our basketball player has jumped approximately a foot in the air.

Pumping Blood

How much energy does it take to pump the blood around the body? What is the power output of the heart? In order to understand the physics of blood circulation, we need to first understand the basic principles of fluid flow in channels or tubes. There are three basic types of fluid flow: frictionless flow, laminar flow, and turbulent flow. Frictionless flow is an ideal case, but there are many situations in real life where viscous friction is negligible. In such cases the approximation of no friction gives adequate results. Another approximation, and one that works well for all three types of fluid flow, is to assume that the fluid is incompressible. Obviously in the case of air in the lungs the fluid, air, is compressible. But for blood in arteries and veins (and for other liquids) the assumption of incompressibility introduces only negligible error.

Bernoulli's Equation

Where friction can be neglected, the flow of an incompressible fluid is described by Bernoulli's equation.

$$P + \rho gh + (1/2)\,\rho v^2 = \text{constant} \qquad\qquad (12\text{-}12)$$

This equation follows from the law of conservation of energy. It simply says that total energy content of a fluid (potential energy plus kinetic energy) is constant. The equation could have been written $E_{\text{Potential}} + E_{\text{Kinetic}} = \text{constant}$.

All three terms in Eq. (12-12) represent a type of energy. The first two terms represent two different forms of potential energy; the third term is the kinetic energy. All are expressed in terms of energy per unit volume of fluid.

Bernoulli's equation tells us that no matter how the flow of a fluid changes, the sum of the potential energy and the kinetic energy remains constant. This allows us to calculate various things. For example, we can calculate the effect of changing the flow velocity on the fluid pressure.

The first term P is the pressure, which is equal to the potential energy of the fluid (due to pressure) per unit volume. We can see this clearly by looking at the units. Standard units of pressure are force per area (newton/meter^2). If we multiply the N/m^2 units by meter per meter (m/m, which doesn't actually change anything), we get newton·meters/meter^3. But a newton·meter is a joule, so the units of pressure (N/m^2) are actually J/m^3, or energy per unit volume.

The second term ρgh is the potential energy due to gravity. The symbol ρ (rho) is the density of the fluid (approximately 1060 kg/m^3 for blood), g is the acceleration due to gravity, and h is the height of the fluid. This term of Bernoulli's equation merely expresses the fact that fluids tend to flow downhill under the influence of gravity. You should be able to show that the units are again energy per unit volume.

The third term $(1/2)\rho v^2$ is the kinetic energy of the fluid (per unit volume). Note that v is the velocity of the fluid. As the velocity of a fluid increases, its kinetic energy increases.

Size of the Vessel

Bernoulli's equation can give us some insight into what happens when a liquid flows from a larger tube to a smaller tube, such as what happens when blood flows from large arteries to smaller arteries or when arteries become narrowed by the buildup of plaque due to arterial disease.

In the remaining discussions of this chapter we will ignore the potential energy due to gravity in Bernoulli's equation. In the situations that we will analyze, the height difference is negligible. The only case where it is significant is in analyzing blood pressure differences in distant parts of the body when a person stands up, for example, differences between the head and chest or between the chest and lower legs. Even then the differences are slight.

Figure 12-2 shows fluid flowing from a larger tube into a smaller tube. The volume of fluid flowing, the flow rate, in each tube is given by the velocity of the fluid times

FIGURE 12-2 • Fluid flow through tubes of two different diameters.

the cross-sectional area of the tube. For an incompressible fluid, the same amount of fluid must flow through each tube. Thus the two flow rates $A \cdot v$ are equal.

$$A_1 v_1 = A_2 v_2 \tag{12-13}$$

Rearranging to give v_2 in terms of v_1, we get

$$v_2 = (A_1/A_2) v_1 \tag{12-14}$$

This tells us that if A_1 is larger than A_2, then v_2 will be larger than v_1. In other words, *as an incompressible fluid flows from a larger tube into a smaller tube, the velocity of the fluid increases.*

What happens to the pressure? Bernoulli's equation tells us that the sum of the terms in Eq. (12-12) is constant. Therefore at any points along the flow the sum of these terms are equal. Taking a point at the wide portion of the tube (subscript 1) and at the narrow portion of the tube (subscript 2), we get

$$P_1 + \rho g h_1 + (1/2) \rho v_1^2 = P_2 + \rho g h_2 + (1/2) \rho v_2^2 \tag{12-15}$$

As mentioned before, we will treat any height differences as negligible, so $h_1 = h_2$. This gives us

$$P_2 - P_1 = (1/2) \rho v_1^2 - (1/2) \rho v_2^2 \tag{12-16}$$

Now if we express v_2 in terms of v_1, as in Eq. (12-14), we get

$$P_2 - P_1 = (1/2) \rho v_1^2 - (1/2) \rho (A_1/A_2)^2 v_1^2$$

or

$$P_2 = P_1 - \frac{1}{2} \rho v_1^2 \left[\left(\frac{A_1}{A_2} \right)^2 - 1 \right] \tag{12-17}$$

This tells us that if A_1 is greater than A_2, then P_2 will be less than P_1. In other words, *as an incompressible fluid flows from a larger tube into a smaller tube, the pressure of the fluid decreases.*

Laminar Flow

Where viscous friction cannot be ignored, the friction tends to slow the flow of the fluid. The viscosity is due to molecular attractions between the fluid molecules, and between the fluid and the walls of the tube or channel in which the fluid is flowing. The fluid closest to the vessel walls moves the slowest. The fluid in the center of the vessel moves the fastest. This type of flow where the velocity of the fluid changes layer by layer, or according to the distance from the vessel walls, is called *laminar* flow.

The flow rate (volume per unit time past a certain point in the channel or vessel) is given by the cross-sectional area times the average fluid velocity. This is similar to the frictionless case except that we must use average velocity because the velocity varies at different points within the cross section of the tube.

$$Q = A \, v_{avg} \tag{12-18}$$

Normally friction tends to slow objects down. But if the average velocity is kept constant, then the result of the friction will be a drop in pressure as the fluid moves along a length of tube. The pressure drop represents a reduction in potential energy as the potential energy (of pressure) is converted into kinetic energy in order to overcome the friction and keep the fluid moving at a constant average velocity. For a viscous (nonfrictionless) fluid, the relationship between the flow rate, the fluid's viscosity, the radius of a tube, and the pressure drop along a given length of tube is given by Poiseuille's Law.

$$Q = \frac{\pi R^4 (P_1 - P_2)}{8 \eta L} \tag{12-19}$$

where R is the radius of the tube, η (the greek letter, eta) is the viscosity of the fluid, L is the length of the tube from point 1 to point 2, and $P_1 - P_2$ is the drop in pressure in going from point 1 to point 2. The interesting thing to note about Poiseuille's law is the dependence on the fourth power of the radius. This means, for example, that doubling the radius will increase the volumetric flow rate 16 times. It also means that for a given (constant) flow rate and viscosity (as we typically have in the case of blood in arteries), a small increase in radius will make the pressure drop significantly less. This is easy to see by rearranging Eq. (12-19) to solve for the pressure drop.

$$P_1 - P_2 = \frac{Q8\eta L}{\pi R^4}$$

(12-20)

Equation (12-20) tells us how much the pressure will drop, due to viscosity, as a fluid flows along a tube with a given Q, η, L, and R (i.e., with a given flow rate, viscosity, length of tubing, and radius). The pressure drop is inversely proportional to the fourth power of the radius. This means that increasing the radius just a little bit will make the pressure drop much smaller (so the effect of viscosity is greatly reduced). It also means that for a liquid such as blood that has a relatively small viscosity to begin with it takes only a moderate radius for the viscous pressure drop to be small enough for Bernoulli's equation to become a very good approximation.

Still Struggling

In the case of blood flowing in the body, the volumetric flow rate, Q, is constant. Yes it goes up and down slightly with the pumping of the heart but on average it is constant along the length of a given blood vessel; that is, the rate of blood flowing into one end of a vessel is the same as is flowing out of the other end. In such a case where the flow rate is constant, viscous friction reduces the pressure. The flow rate is the same at both ends of the vessel, but the pressure is different. However, if this pressure change is very small, then the situation can be approximated by a frictionless fluid (Bernoulli's equation). Since the pressure drop is inversely proportional to the radius to the fourth power, a small change in the radius can make a big change to the pressure drop. A slight increase to the radius can make the pressure drop small and insignificant so that Bernoulli's equation can be applied. Conversely, a decrease in the radius can cause the pressure drop to be large enough to cause significant negative health effects.

Turbulent Flow

As the velocity of fluid flow increases, a point is reached where the flow is no longer smooth and laminar. Instead *eddies* are formed. These are small currents of fluid that flow backward, opposite the direction of the overall flow. See Fig. 12-3.

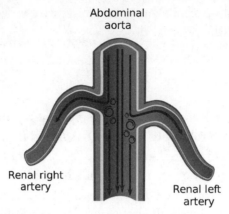

FIGURE 12-3 · Turbulent fluid flow with eddies (circular currents).

The velocity at which flow changes from laminar to turbulent is called the *critical flow velocity*. The critical flow velocity is proportional to the viscosity, and to a property of the fluid known as the Reynolds number. It is inversely proportional to the fluid density and to the radius of the tube.

$$v_c = \frac{\Re\eta}{2\rho R} \qquad (12\text{-}21)$$

From Eq. (12-21) we see that increased viscosity and increased Reynolds number ($\Re$) increase the critical velocity; this makes it *less* likely at a given velocity that the flow will be turbulent, whereas increasing the radius or the fluid density increases the likelihood of turbulent flow.

Still Struggling

A higher critical velocity means the fluid must flow faster in order for turbulent flow to occur. Therefore, high critical velocity means more velocities at which flow is *not* turbulent, and so less possibility for turbulent flow, Thus Eq. (12-21) tells us that increasing viscosity makes it harder for turbulence to occur (due to a higher critical velocity). Whereas increasing radius makes it easier for turbulence to occur (due to a lower critical velocity). In regard to blood this means that if turbulence occurs, it is most likely to occur in the larger blood vessels and less likely in the smaller ones.

PROBLEM 12-2

The volumetric flow rate of blood in a person at rest is about 5 L per minute. If the aorta (the central artery carrying blood from the heart) has a diameter of 2 cm, what is the kinetic energy of blood flowing through the aorta?

SOLUTION

From Eq. (12-18) we can calculate the average velocity of blood through the aorta. A liter is 1000 cm³. The cross-sectional area of a tube with a diameter of 2 cm (radius 1 cm or 10^{-2} m) is

$$A = \pi r^2 = (3.1416)\,(10^{-2}\,\text{m})^2 = 3.1416 \times 10^{-4}\,\text{m}^2$$

So the velocity is

$$V_{avg} = \frac{Q}{A} = \frac{(5000\,\text{cm}^3/\text{min})\,(10^{-6}\,\text{m}^3/\text{cm}^3)\,(1\,\text{min}/60\,\text{s})}{3.1416 \times 10^{-4}\,\text{m}^2} = 0.2653\,\text{m/s}$$

The kinetic energy is given by the last term in Bernoulli's equation [Eq. (12-12)].

$$E_{\text{Kinetic}} = (1/2)\,\rho v^2 = (0.5)\,(1060\,\text{kg/m}^3)\,(0.2653\,\text{m/s})^2 = 140.6\,\text{J/m}^3$$

PROBLEM 12-3

The volumetric flow rate of blood in a person at rest is about 5 L per minute. The beginning of the aorta (the part closest to the heart) has a diameter of 3 cm, but the end of the aorta in the lower abdomen has a diameter of only 1.75 cm. How much faster is the blood traveling at the end of the aorta than it is at the beginning?

SOLUTION

Five liters per minute is (5000 cm³/min) (10^{-6} m³/cm³) (1 min/60 s) = 8.333×10^{-5} m³/s. We can use Eq. (12-18) to calculate the average velocity at each end of the aorta. At the beginning where the diameter is 3 cm, the radius is 0.015 m, so

$$V_{avg} = \frac{Q}{A} = \frac{Q}{\pi r^2} = \frac{8.333 \times 10^{-5}\,\text{m}^3/\text{s}}{(3.1416)\,(0.015\,\text{m})^2} = 0.1179\,\text{m/s}$$

At the end of the aorta, where the diameter is 1.75 cm, the radius is 0.00875 m, so

$$v_{avg} = \frac{Q}{A} = \frac{Q}{\pi r^2} = \frac{8.333 \times 10^{-5} \, m^3/s}{(3.1416)(0.00875 \, m)^2} = 0.3464 \, m/s$$

The answer is that the blood is traveling with a velocity 2.94 times faster at the end of the aorta compared with the beginning.

Arteriosclerosis

Arteriosclerosis is a disease characterized by a hardening on the arteries. The most common form of arteriosclerosis, called *atherosclerosis*, involves a buildup of fatty deposits on the artery walls which later becomes hardened with calcium deposits. Aside from the loss of elasticity in the arterial walls, there is a decrease in arterial radius. If we take Eq. (12-18) and put it in the form $v_{avg} = Q/(\pi r^2)$, we clearly see the inverse square relationship between average velocity and radius. This inverse square relationship means a 30% decrease in arterial radius will double the velocity of the blood. And a 50% decrease in arterial radius will quadruple the velocity of blood. The kinetic energy of the blood increases 16-fold (since kinetic energy is proportional to the velocity squared).

This significant increase in kinetic energy occurs through a proportional decrease in blood pressure. We can see from Eq. (12-20) that the pressure drop due to viscous friction is inversely proportional to the radius to the fourth power. The result is that a 50% decrease in radius will create a pressure drop along the length of a blood vessel that is 16 times larger than it would be otherwise.

All of this combined can be very detrimental to one's health. One of the worst effects is the possibility of turbulent blood flow in the arteries. While a decrease in radius will increase the critical flow velocity, making turbulent flow less likely [Eq. (12-21)], it does so only to the first power of the radius; whereas the same decrease in radius increases the velocity by the square of the radius. So the velocity increases much faster than the critical flow velocity, making it much more likely that the critical flow velocity will be exceeded and smooth laminar blood flow will be replaced by turbulent flow. This is especially dangerous during strenuous activity when the volumetric blood flow and average velocity can be as much as 5 to 6 times larger than when at rest. Once turbulent flow commences, the heart has to work a lot harder to maintain the same flow rate. Turbulent flow is also dangerous because it puts more stress on the walls of blood vessels, which can weaken the arterial walls or can dislodge some of the deposits. These dislodged deposits then float through the bloodstream and can cause a blockage elsewhere in the body.

Hummingbird Hovering

Many birds are able to hover, to fly in a relatively fixed position, not going up or down, left or right, forward or backward. Many large birds, for example seagulls, can hover for brief periods of time by adjusting their wings to ride the breeze like a kite. This of course depends on the consistency of the wind, and the bird's ability to rapidly adjust to changes in wind currents. The dynamic forces involved are rather intricate and so we will not take the time to examine them here. However, hummingbirds can hover for extended periods of time, with no breeze at all, by flapping their wings very rapidly.

Let's examine what happens when a hummingbird hovers. In order to hover, an organism must generate an upward force that is, on average, equal to its own weight. The upward force balances out the force of gravity (the organism's weight) so that no upward or downward acceleration exists. For this discussion we will ignore forces in directions other than vertical. But you should know and consider that what follows is simplified. In practice the movements and forces necessary for extended hovering in birds are quite complex, since the bird must also control its horizontal motion (especially if it is trying to drink nectar from a flower while hovering). Still, since motion in general can be mathematically resolved into each of the three dimensions, and each dimension dealt with independently, the discussion that follows regarding the vertical dimension is correct for that dimension. We're just not going to complicate matters by including the horizontal forces and motions in our treatment.

The lifting force needed for hovering is not constant, but varies with the movements of the wings. This is why we said that the force must be *on average* equal to the weight of the bird. As long as the *average* force equals the weight, then the bird can hover, provided the fluctuations in the upward force are rapid. This is because every time the upward force is less than the weight, the bird will accelerate downward (fall). And every time the upward force is greater than the weight, the bird will accelerate upward (rise). If these fluctuations are relatively small and rapid, then the rising and falling of the bird will be minimal, or even negligible. But if the changes in lift are large or slow, then bird will end up bobbing up and down in the air instead of hovering.

FIGURE 12-4 · Hummingbird horizontal flight. (*Courtesy of Wikimedia Commons.*)

Average Lift

Let's make an approximation, or simplifying assumption, that the lifting force is constant for the duration of each stroke of the wings. Then the only fluctuation in force that we need to deal with is the difference between the force of the downward stroke and the force of the upward stroke. With this approximation we have only two different lift forces that average over time to compensate for the weight of the bird.

The upward lift comes from the reactive force of air pushing up on the wing when the wing pushes down on the air (Newton's third law). Depending on the angle of the wing, it is possible for lift to be generated on both the downstroke and the upstroke. Figure 12-4 shows a diagram of a hummingbird in horizontal flight. The beating of the wings provides both lift and forward thrust. Compare this with Fig. 12-5 which illustrates sustained hovering flight. When hovering, the hummingbird orients itself so that the downstroke is somewhat of a forward stroke (relative to the ground), and the upstroke is somewhat of a backward stroke. This orientation enables both strokes to provide lift. (We still

FIGURE 12-5 · Hummingbird sustained hovering flight. (*Courtesy of Wikimedia Commons.*)

call them downstroke and upstroke, because relative to the bird's anatomy the strokes are down and up.)

There are two extreme cases we can look at in terms of the lifting force generated by the beating of the wings. At one extreme, both strokes equally generate lift, and so the lifting force is essentially constant. At the other extreme, lifting force is generated only on the downstroke, and no vertical force is generated by the upstroke. Let's examine this latter case in more detail.

When lifting force is generated only on the downstroke, the force alternates with the beating of the wings, between a positive value of upward lift and zero. In such a case the bird will fall slightly during the upstroke (due to gravity) and rise slightly during the downstroke (due to upward lift that is larger than the weight of the bird). This size of the up and down motion will depend on the amount of time for a wing stroke. On the downward stroke, the lifting force accelerates the bird upward for the duration of the stroke. On the upward stroke, with no lifting force, gravity accelerates the bird downward, again for the duration of the stroke. Depending on the species, a hummingbird can beat its wings anywhere from 20 to 100 times per second.

If we assume an equal amount of time is spent in each stroke (up or down), then we can calculate the amount of force required on the downstroke. The average lifting force must equal the weight of the bird. Since the upstroke exerts zero force, the downstroke must therefore exert a force equal to 2 times the weight of the bird in order for the average of the two numbers to equal the weight of the bird. (If either the down- or upstroke lasted longer than the other, then we would calculate a weighted average, weighted according to the percentage of time of one wing beat cycle spent in each stroke.)

PROBLEM 12-4

Assume a hummingbird beats its wings (one complete cycle of downstroke and upstroke) 20 times per second. Assume also that the time for each stroke is exactly half a beat cycle. Assume that the only vertical force on the bird during the upstroke is due to gravity, and the only vertical forces during the downstroke are gravity and a lift force equal to twice the weight of the bird. Calculate the height of the up and down motion of the bird while hovering.

✔ SOLUTION

We only need to calculate the downward motion. The upward motion is exactly the same only in the opposite direction. This is because during the downward motion the only force is the weight of the bird. During the upward motion there are two forces: an upward force equal to twice the weight of the bird, and the downward force equal to the weight of the bird. The net force is therefore equal to the weight of the bird, but in the upward direction.

Since $F = ma$ (and the mass is constant) and since the upward and downward forces are equal in magnitude (just occurring at different times in different directions), then the accelerations are also equal in magnitude. Therefore we can use the acceleration due to gravity to calculate how much the bird drops during the upstroke, and assume that during the downstroke the bird rises by the same amount.

We know the acceleration due to gravity is -9.807 m/s^2. Each stroke is half a cycle, and there are 20 cycles (wing beats) per second. So the length of time for one stroke is 1/2 of 1/20 of a second, or 1/40 of a second.

During the upstroke, the bird accelerates -9.807 m/s^2 for 1/40 of a second. The distance the bird drops during this time is equal to the average velocity multiplied by the time [from Eq. (12-6)].

$$d = v_{avg}\, t \tag{12-22}$$

Since the acceleration is constant, the average velocity is just the average of the initial and final velocities [Eq. (12-5)]. We can assume the initial velocity is zero (at the beginning of the upstroke). Therefore the average velocity is just $v_2/2$.

$$d = (v_2/2)\, t \tag{12-23}$$

We get the final velocity v_2 from Eq. (12-4) by setting $v_1 = 0$; this gives $v_2 = a\, t$.

Substituting this into Eq. (12-23) gives us $d = (a\, t/2)\, t$, or

$$d = (1/2)\, at^2 \tag{12-24}$$

Therefore the distance the hummingbird drops during the upstroke is

$$d = (0.5)\,(-9.807 \text{ m/s}^2)\,(0.025 \text{ s})^2 = -0.00306 \text{ m}$$

At 20 beats per second the hummingbird rises and falls only 3 mm while hovering (and it does so 20 times per second).

Hummingbird Wing Lift

Most birds generate lift only on their downstroke. However, most birds are not able to hover. Many flying insects can hover, and many insects are able to rapidly twist or angle their wings with each stroke so that they generate lift equally on both the upstroke and downstroke. Hummingbirds however fall somewhere in between these two extremes, or at least the one species of hummingbird that has been studied thoroughly in this regard.

Recent research examined hummingbird sustained hovering flight using *digital imaging particle velocimetry (DPIV)*. DPIV is a technique that involves suspending small tracer particles (10 to 100 μm in size) in a fluid, such as air. The particles are illuminated with laser light, and high-speed digital video cameras record the location of the particles as many as 500 times per second. Computer analysis can then be used to calculate the velocity of the particles and the force required to accelerate them to that velocity. Researchers observed rufous hummingbirds hovering in a variable-speed wind tunnel, and used DPIV to analyze air currents generated by the hummingbirds' wings. They discovered that, while hovering, hummingbirds generate 75% of their lift during the downstroke, and 25% of their lift during the upstroke. In other words both strokes generate a lifting force, but the downstroke generated 3 times as much lift as the upstroke.

We can use this information to calculate the force generated by each stroke. Let's call $LF_{downstroke}$ the lifting force generated by the downstroke and $LF_{upstroke}$ the lifting force generated by the upstroke. We know the average lifting force must equal the weight of the hummingbird. Assuming that the downstroke and upstroke take equal amounts of time, the average is

$$(LF_{downstroke} + LF_{upstroke})/2 = \text{weight} \qquad (12\text{-}25)$$

We also know that the downstroke generates 3 times as much lift as the upstroke.

$$LF_{downstroke} = 3LF_{upstroke}$$

Substituting this into Eq. (12-25) gives

$$3\,LF_{upstroke} + LF_{upstroke} = 2\,\text{weight}$$

or

$$4\,LF_{upstroke} = 2\,\text{weight}$$

or

$$LF_{upstroke} = (2/4) \text{ weight} = (1/2) \text{ weight} \qquad (12\text{-}26)$$

Thus the lifting force of the upstroke 0.5 times the weight. Since the average must equal the weight, then the lifting force of the downstroke must be 1.5 times the weight.

 PROBLEM 12-5

Making all the same assumptions as in Prob. 12-4 (20 wing beats per second, etc.) except now assuming that 75% of the lift is generated by the downstroke and 25% of the lift is generated by the upstroke, calculate the distance that the hummingbird moves up and down with each wing beat.

 SOLUTION

This problem is almost identical to Prob. 12-4, except that some of the lift is provided by the upstroke. This means that the force, and therefore the acceleration, is different. On the upstroke the wing provides a lift equal to 0.5 the weight of the bird. Gravity provides a downward force equal to the weight of the bird. Therefore the net force on the hummingbird during the upstroke is a downward force equal to half the weight of the bird. The bird's mass, however, has not changed ($F = ma$); therefore since the downward force is only half the weight of the bird (and the mass has not changed), then the downward acceleration is only half the acceleration due to gravity, $(1/2)(-9.807 \text{ m/s}^2) = -4.903 \text{ m/s}^2$.

From Eq. (12-24) we get

$$d = (0.5)(-4.903 \text{ m/s}^2)(0.025 \text{ s})^2 = -0.00153 \text{ m}$$

By providing only 25% of the upward lift during the upstroke, the hummingbird has reduced its up and down motion from 3 mm to only 1.5 mm (compared to the case where all of the upward lift is provided by the downstroke).

QUIZ

Refer to the text in this chapter if necessary. Answers are in the back of the book.

1. Which of the following is *not* usually considered part of physiological and anatomical biophysics?

    A. Investigation of cellular uptake of nutrients in the cells lining the small intestine

    B. Fluid dynamics as it relates dolphins swimming in the ocean

    C. Calculation of the muscular forces involved in a golf swing

    D. Measurement of the minimal energy in sound waves necessary for hearing

2. When an animal jumps into the air, as soon as it leaves the ground there is no longer a force pushing the animal up into the air. When this happens,

    A. there is no longer any upward velocity.

    B. its upward velocity remains constant.

    C. there isn't any acceleration.

    D. its upward velocity begins to decelerate to zero.

3. The force that lifts a jumping organism into the air comes from

    A. helium.

    B. the ground pushing up on the organism.

    C. the organism's muscles pushing the organism into the air.

    D. the organism's wings.

4. Blood flow in arteries and veins is normally

    A. red.

    B. laminar.

    C. frictionless.

    D. turbulent.

5. For a person at rest, the blood pressure in the aorta is 120 torr and the velocity of the blood is 25 cm/s. The person then goes for a brisk walk raising the blood velocity to 100 cm/s. Using Bernoulli's equation and assuming the height of the blood remains constant, calculate the blood pressure in the aorta during the brisk walk.

    Hint: Convert everything to standard units; for example, 1 cm = 0.01 m. The standard unit for pressure is pascal (Pa, although blood pressure is usually reported in torr). 1 torr = 133.32 Pa (1 Pa = 1 kg$\cdot$m$^{-1}\cdot$s^{-2}).

    A. 116 torr

    B. 124 torr

    C. 15501 torr

    D. 133 Pa

6. The velocity of blood is about 125 cm/s in the aorta while running. What is the blood's kinetic energy?

    A. 125 J/m^3

    B. 770 J/m^3

C. 828 J/m^3

D. 1656 J/m^3

7. **For a hummingbird to hover, the lifting force generated by the beating of its wings, averaged over the time of one wing cycle, must be**

A. twice the weight of the bird.

B. converted to air pressure to keep the bird in the air.

C. at least 20 beats per second.

D. equal to the weight of the bird.

8. **A certain species of hummingbird, while hovering, exerts 75% of its lifting force on the downstroke of its wings and 25% of the lifting force on the upstroke. In order to keep the up and down motion of the bird to less than 0.5 mm, how many times per second will the bird have to beat its wings?**

Hint: Use Eq. (12-24) and solve for *t*, the time for one beat cycle.

A. 70 times per second

B. 50 times per second

C. 35 times per second

D. 20 times per second

9. **Turbulent blood flow can occur if the velocity of blood is too fast. It also**

A. is proportional to the amount of laminar blood flow.

B. can lead to deposits on the walls of arteries.

C. occurs if the critical flow velocity is too large.

D. makes the heart work harder.

10. **A dolphin jumps out of the water. At the moment the dolphin leaves the water, it has an upward velocity of 7 m per second. How high above the water will the dolphin go before it begins to descend back down to the water?**

Hint: Use Eq. (12-9).

A. 2.5 m

B. 2.2 m

C. 2.1 m

D. 0.4 m

Final Exam

Select the best answer for the following multiple-choice questions. The answers are located in the next section.

1. **What does it mean if the Gibbs energy change for a process is positive?**

    A. The process is favorable.

    B. The process is reversible.

    C. The reverse process is favorable.

    D. The partition function is more difficult to calculate.

    E. The entropy is decreasing.

2. **What is the cell nucleus?**

    A. A membrane bound organelle found in eukaryotes

    B. The cell's genetic material

    C. A set of nucleotides found at the center of a chromosome

    D. The center of a neutron star

    E. A membrane that surrounds the nuclear envelope

3. **DNA is a**

    A. heteropolymer.

    B. homopolymer.

    C. peptide polymer.

    D. ribonucleotide polymer.

    E. hydrocarbon polymer.

4. An instrument that characterizes molecules according to their molecular weight by ionizing them and passing them through a magnetic field in a vacuum is called

   A. an ionization magneto spectrometer.

   B. a molecular scale.

   C. molecular tweezers.

   D. a mass calorimeter.

   E. a mass spectrometer.

5. How many *distributions* are possible for given a Boltzmann distribution for a given number of molecules and a specified amount of energy?

   A. Infinite number,

   B. 1.

   C. 2.

   D. 10^{23}.

   E. It depends on how much energy is available.

6. In biophysics, when the molecules making up a specific quaternary structure are all that is required to build that structure and there is no need for energy input or catalysis, we call this

   A. prefab construction.

   B. quaternary binding.

   C. self-assembly.

   D. self-aggregation.

   E. energy-free assembly.

7. Biomechanics includes

   A. electrophysiology.

   B. sensory biophysics.

   C. animal motion and forces.

   D. anatomical steric forces.

   E. allosterics.

8. Which of the following is *not* true about proteins?

   A. Proteins are polymers of amino acids.

   B. Polypeptide chains fold up into a specific shape necessary for the protein's function.

C. Transfer RNAs carry amino acids to where they are needed for protein production.

D. Peptide bonds hold base pairs together.

E. In eukaryotes, most protein synthesis takes place on ribosomes, whereas in prokaryotes most protein synthesis takes place in the cytoplasm, near the cell membrane.

9. **Scanning differential calorimetry**

A. measures heat capacity as a function of temperature.

B. can be used to obtain direct thermodynamic data for phase transitions or temperature-induced conformational changes.

C. provides structural information from heat capacity data.

D. A and B

E. B and C

10. **Passive transport refers to**

S1. **cells floating passively in solution.**

S2. **small molecules diffusing through a membrane.**

S3. **transport through a membrane that does not require a transport protein.**

S4. **ions passing through a tunnel created by a transport protein.**

A. S1 and S2

B. S2 and S3

C. S3 and S4

D. S1 and S3

E. S2 and S4

11. **Given the following two expressions for the partition function**

$$Z_A = 1 + \sum_{i=1}^{L_{MAX}} \omega_i e^{-\frac{\Delta E_i}{kT}} \qquad \text{and} \qquad Z_B = \sum_{i=1}^{L_{MAX}} \omega_i e^{-\frac{E_i}{kT}}$$

which of the following statements is true?

A. The partition function Z_A contains a greater amount degeneracy than Z_B.

B. The partition function Z_B contains a greater amount degeneracy than Z_A.

C. The partition function Z_A is expressed relative to a reference state E_0 with a relative probability of $1/Z_A$.

D. The partition function Z_B is expressed relative to a reference state E_0 with a relative probability of $1/Z_B$.

E. Z_A is not the correct expression for a partition function; Z_B is.

12. **Which of the following are true regarding X-ray crystallography?**

S1. X-ray crystallography provides very precise, high-resolution structural information for molecules.

S2. X-ray crystallography is also called *X-ray diffraction*.

S3. For X-ray crystallography to work, the molecules must be in a regular, repeating arrangement (also known as a *crystal*).

S4. X-ray diffraction images were used to determine that DNA is a double helix.

A. S1, S2, and S3

B. S2, S3, and S4

C. S1, S3, and S4

D. S1, S2, and S4

E. S1, S2, S3, and S4

13. **A ligand is**

A. connective tissue that holds bones together.

B. connective tissue that ties muscles to bones.

C. an element in the periodic table.

D. a smaller molecule or atom that binds to a larger molecule.

E. a cell structure that produces the enzyme called *ligase*.

14. **A biophysicist wants to use a two-state approximation to model the conformational state of a solution of lipids. That is, we assume that the lipids can exist in only two energy states: Assuming a Boltzmann distribution and using $E_1 = 2.5 \times 10^{-20}$ J and $E_2 = 3.0 \times 10^{-20}$ J, what is the percentage of lipids in each energy level at 310 K?**

A. $F_1 = 77.4\%$ and $F_2 = 22.6\%$.

B. $F_1 = 76.3\%$ and $F_2 = 73.7\%$.

C. $F_1 = 77.8\%$ and $F_2 = 22.2\%$.

D. $F_1 = 76.3\%$ and $F_2 = 23.7\%$.

E. The answer cannot be determined from the information given.

15. A molecule made from only carbon, hydrogen, and oxygen is called what?

    A. Carbonated water

    B. Hydrocarbon

    C. Lipid

    D. Carbohydrate

    E. Carboxylic acid

16. In which of the following interactions is the internal energy proportional to $1/r$, where r is the distance between the entities?

    A. Charge-dipole interactions

    B. Dipole-induced dipole interactions

    C. Charge-charge interactions

    D. A and B

    E. B and C

17. Decreasing temperature will make a process more likely to occur if the process

    S1. releases energy while decreasing entropy.

    S2. releases energy while increasing entropy.

    S3. absorbs energy while decreasing entropy.

    S4. absorbs energy while increasing entropy.

    A. S1 and S2

    B. S1 and S3

    C. S1 and S4

    D. S2 and S3

    E. S3 and S4

18. If the *superhelix density* is defined as the absolute value of the writhe divided by 1/10 the number of base pairs, what is the superhelix density of a 6000-base-pair closed duplex DNA with linking number of 565 and a twist of 595?

    A. 0.05

    B. 0.99

    C. 0.09

    D. 0.10

    E. 0.30

19. **In the gel state**

    A. small holes in the lipid bilayer are easily repaired.

    B. the fluid mosaic is decrystallized.

    C. the fluid mosaic contains cholesterol.

    D. small holes in the lipid bilayer are filled with gel.

    E. lipid movement is restricted by dispersion forces.

20. **Use the following table to calculate the length of a 5000-base-pair DNA molecule in the A-DNA conformation and in the B-DNA conformation.**

Geometric Parameter	B-DNA	A-DNA	Z-DNA
Pitch of helix (length of one turn)	33.2 Å	24.6 Å	45.6 Å
Width of helix	23.7 Å	25.5 Å	18.4 Å
Average base pairs per turn of helix	10.0	10.7	12

    A. A-DNA length $= 1.15 \times 10^4$ Å, B-DNA length $= 1.66 \times 10^4$ Å.

    B. A-DNA length $= 1.66 \times 10^4$ Å, B-DNA length $= 1.15 \times 10^4$ Å.

    C. A-DNA length $= 1.27 \times 10^3$ Å, B-DNA length $= 1.66 \times 10^3$ Å.

    D. A-DNA length $= 8.66 \times 10^4$ Å, B-DNA length $= 6.11 \times 10^4$ Å.

    E. The length cannot be calculated from the information provided.

21. **Statistical mechanics falls into which division of biophysics?**

    A. Molecular and subcellular biophysics

    B. Physiological biophysics

    C. Environmental biophysics

    D. All of the above

    E. None of the above

22. **What is a nucleic acid palindrome?**

    A. A nucleotide sequence that can form a stem-and-loop or cruciform structure

    B. A nucleotide sequence that reads the same forward or backward

    C. A protein that binds to certain DNA sequences

D. A dome-shaped DNA secondary structure

E. None of the above

23. **Fill in the blanks: Two-chain phospholipids tend to form _____ whereas single-chain phospholipids form _____.**

A. micelles, liposomes

B. lysozomes, micelles

C. liposomes, cholesterol

D. liposomes, micelles

E. membranes, vesicles

24. **Which of the following statements is *most* true?**

A. A *voltage clamp* is a technique used in electrophysiology to measure and characterize voltage differences in cells, particularly in excitable tissue cells such as neurons (nerve cells).

B. A *voltage clamp* is a technique used in electrophysiology to measure and characterize electrical currents in cells, particularly in excitable tissue cells such as neurons (nerve cells).

C. A *voltage clamp* is a technique used in conjunction with electrophoresis, to measure and characterize voltage differences among molecules.

D. The *current clamp* technique is similar to the voltage clamp, except that the electrode feedback mechanism allows the current across the membrane to vary while keeping a constant voltage.

E. None of the above

25. **DNA owes its helical shape to**

A. Watson and Crick.

B. Rosalind Franklin.

C. hydrogen bonds between base pairs.

D. aromatic base stacking.

E. the helical phosphate backbone.

26. **The Boltzmann distribution**

A. is the most probable distribution for any significant number of molecules.

B. is not the only distribution that occurs.

C. has units of joules per molecule.

D. A and B

E. B and C

27. Which property of cdDNA is a topological invariant?

  A. Lk

  B. Tw

  C. Wr

  D. A and B

  E. B and C

28. Secondary active transport is called *secondary* because its source of energy is

  A. an ionic current that was generated by a gated ion channel.

  B. a temperature gradient that was generated by a primary active transport.

  C. a phospholipid gradient that was generated by secondary photosynthesis.

  D. a concentration gradient that was generated by a primary active transport.

  E. the second of three pyrophosphate bonds in ATP.

29. The cell cycle describes

  A. how certain cells with flagella swim in circles to surround their food.

  B. the cycle of food entering the cell and being digested, and energy from that food used to create ATP, which is used to actively transport more food into the cell (and then the cycle repeats).

  C. how cells grow and then divide in daughter cells, which grow and then divide, and so on.

  D. any of a variety of cyclical reactions in the cell, for example, the sodium-potassium pump, which has a cycle of repeatedly pumping sodium out of the cell and potassium into the cell.

  E. There's no such thing as the cell cycle. You just made that up.

30. Which amino acids tend to be on the outside of a folded protein that is floating around in a cell?

  A. Nonpolar

  B. Uncharged polar

C. Charged

D. A and B

E. B and C

31. **What are the three major divisions of biophysics when categorizing the branches of biophysics according to size of what is being studied?**

    A. Small, medium and large biophysics

    B. Microscopic, tissue, and organism biophysics

    C. Subcellular, physiological, and environmental biophysics

    D. Subcellular, structural, and macromolecular biophysics

32. **Fill in the blanks: The ultracentrifuge operates on the principle of _____ and can be used to determine the relative _____ of molecules.**

    A. electrophoresis, size

    B. sedimentation, strengths

    C. centrifugation, elasticity

    D. gravity, compactness

    E. sedimentation, compactness

33. **If we double the number of segments in a freely jointed chain, what is the effect on the root-mean-square end-to-end distance?**

    A. Increase by a factor of 2

    B. Increase by a factor of $\sqrt{2}$

    C. Decrease by a factor of 2

    D. Decrease by a factor of $\sqrt{2}$

    E. Remain constant because the average length of each segment will balance out the change

34. **What do disulfide bonds do?**

    A. Provide a favorable Gibbs energy change for bringing together otherwise distant cysteine residues in a folded protein.

    B. Add rigidity to the folded protein structure.

    C. Form ionic bonds between two charged, sulfur-containing amino acids.

D. A and B

E. B and C

35. **How many *arrangements* are possible given a Boltzmann distribution for a specific number of molecules and a specific amount of energy?**

   A. It depends on how much energy is available.

   B. The average of all possible distributions.

   C. The maximum of all possible distributions.

   D. One.

   E. The answer cannot be determined on the basis of the information given.

36. **An amino acid is a molecule with what functional groups?**

   A. An amine and hydrochloric acid

   B. A hydroxyl and an amine

   C. A protein and a polypeptide

   D. A carboxyl and an amine

   E. A hydrogen bond and a dipole

37. **Which of the following is *not* a membrane-bound organelle?**

   A. Mitochondrion

   B. Chloroplast

   C. Vacuole

   D. Ribosome

   E. Nucleus

38. **In an alpha helix**

   A. the backbone is on the inside of the helix.

   B. the side chains have dipoles.

   C. the dipoles are every fourth residue.

   D. the hydrogen bonds occur on the alpha residues.

   E. the base pairs are tilted slightly.

39. **Absorption of light is commonly used to measure**

   S1. conformational transitions.

   S2. voltage across a membrane.

   S3. ligand binding.

S4. heat capacity.

    A. S1 and S2

    B. S2 and S3

    C. S3 and S

    D. S1 and S3

    E. S2 and S4

40. **Antiparallel strands are found in**

    A. mRNA.

    B. phospholipid hydrocarbon chains.

    C. beta sheets.

    D. alpha helices.

    E. lipid bilayers.

41. **Cooperativity is**

    A. the occurrence of separate events together in a nonindependent manner.

    B. several proteins cooperating to form a quaternary structure.

    C. the alpha helix and the beta sheet working together to fold a protein.

    D. a series of independent two-state processes.

    E. two cell membranes fusing together.

42. **Sedimentation describes**

    A. the process of dead organisms forming layers on the ocean floor.

    B. a type of rock with fossils in it.

    C. the motion of particles in a fluid under the application of a force.

    D. an analytical ultracentrifuge with special optical devices and sensors.

    E. the release of waste from plant cells into the soil.

43. **An analytical ultracentrifuge contains special optical devices and sensors that can track the movement of molecules as they are being centrifuged. From this we can learn the**

    A. equations of sedimentation.

    B. size and approximate shape of molecules.

    C. Gibbs energy change for moving a molecule through a fluid.

    D. electric field strength required to ionize large proteins.

    E. sequence of DNA.

44. **A cell is**

    A. a terrorist organization.
    B. a type of phone.
    C. the basic unit of life.
    D. an egg.
    E. a virus.

45. **Which of the following items apply to prokaryotes?**

    S1. Typically larger
    S2. Typically smaller
    S3. Membrane bound organelles
    S4. Shorter cell cycle (rapid division)
    S5. Simpler biochemical reactions
    A. S1 and S3
    B. S2 and S4
    C. S1 and S4
    D. S2 and S3
    E. S2, S4, and S5

46. **Ribosomes are**

    A. quaternary structures that contain protein and RNA.
    B. quaternary structures that contain protein and DNA.
    C. tertiary structures that synthesize polypeptides.
    D. tertiary structures that contain protein, DNA, and RNA.
    E. none of the above

47. **What are the major forces that hold membranes together?**

    A. Dipole induction and phospholipid stacking
    B. Dispersion forces and hydrogen bonds
    C. Hydrogen bonds and hydrophobic interactions
    D. Hydrophobic interactions and dispersion forces
    E. Peptide bonds and hydrophilic interactions

48. **Which of the following statements is true?**

    S1. Superhelical DNA requires a force to hold the superhelix in place.

S2. Circular DNA is always superhelical.

S3. A superhelix is a quaternary structure.

S4. Energy stored in the superhelix can be used to unwind the DNA.

A. S1 and S2

B. S1 and S3

C. S1 and S4

D. S2 and S3

E. S2 and S4

49. Which of the following statements about Φ,Ψ contour diagrams are true?

S1. Φ and Ψ are angles for the bonds in the polypeptide backbone that can rotate.

S2. Contours are plotted for values of Φ and Ψ for which the potential energy is favorable.

S3. Steric hindrances disallow certain values of Φ and Ψ.

A. S1 and S2

B. S2 and S3

C. S1 and S3

D. S1, S2, and S3

E. None of the above

50. The sedimentation force in gel electrophoresis comes from

A. the sedimentation coefficient.

B. gravity.

C. an electric field.

D. all of the above

E. none of the above

51. The process of converting the nucleotide sequence of an mRNA molecule into an amino acid sequence in a polypeptide is called what?

A. Transport protein synthesis

B. Transcription

C. Translation

D. Transformation

E. mRNA to amino acid conversion

52. Molecules on the surface of the cell membrane that bind specific out-side molecules and thereby signal the cell to do something are called

    A. signal proteins.

    B. ribosomes.

    C. receptors.

    D. antennae.

    E. cell walls.

53. What do the Golgi apparatus and the endoplasmic reticulum have in common?

    A. They are both folded membranes that provide surface area for bio-chemical reactions.

    B. They are both part of the endomembrane system.

    C. They are both involved in protein synthesis from amino acids.

    D. A and B

    E. B and C

54. A major role of lipids is what?

    A. Muscle contraction

    B. Phosphorylation of DNA

    C. Structure of phospholipids

    D. Structure of membranes

    E. Digestion

55. Carbon atoms usually form how many covalent bonds?

    A. 2

    B. 4

    C. 6

    D. 8

    E. 10

56. DNA and RNA are examples of

    A. nucleic acids.

    B. nucleotides.

    C. proteins.

D. lipids.

E. topoisomerases.

57. **A beta sheet is**

    A. a common tertiary structure in proteins.

    B. two polypeptide strands with their backbones hydrogen bonded together.

    C. a thin covering on the outside of the cell wall in some organisms.

    D. a common secondary structure in membranes.

    E. a series of flat aromatic molecules connected side to side.

58. **What source of energy does primary active transport use to move molecules through a membrane?**

    A. Optical tweezers

    B. Internal pressure

    C. Pyrophosphate bonds

    D. Concentration gradient

    E. Gibbs energy

59. **Polar bonds are**

    A. ionic bonds.

    B. covalent bonds in which the electrons are shared unevenly.

    C. stronger than hydrogen bonds.

    D. A and B

    E. B and C

60. **The internal energy that results from a charge-dipole interaction**

    A. depends on the angle between the dipole axis and a straight line between the dipole and the charge.

    B. decreases faster than it does for a charge-charge interaction, as the distance between the charge and the dipole increases.

    C. increases faster than it does for a charge-charge interaction, as the distance between the charge and the dipole increases.

    D. A and C

    E. A and B

61. Forces that arise from synchronized fluctuating charge distributions are classified as

    A. van der Waals forces.

    B. dispersion forces.

    C. synchronization forces.

    D. A and B

    E. B and C

62. Stacking interactions are *most* important to which class of biomolecule?

    A. Proteins

    B. Lipids

    C. Nucleic acids

    D. Amino acids

    E. Tryptophan, tyrosine, and phenylalanine

63. The helical shape of DNA is due primarily to

    A. aromatic side chains.

    B. base stacking.

    C. hydrogen bonds.

    D. van der Waals constant.

64. Proteins are involved in

    A. DNA synthesis.

    B. enzymatic processes.

    C. lipid metabolism.

    D. all of the above

    E. none of the above

65. The Gibbs energy change of bringing a hydrophobic molecule in contact with water is unfavorable primarily due to a

    A. large decrease in the hydrophobic molecule's entropy.

    B. large increase in the hydrophobic molecule's entropy.

    C. large decrease in the water's entropy.

    D. large increase in the water's entropy.

    E. positive Gibbs energy change.

66. **In which of the following does the hydrophobic effect play a role?**

    A. Hydration of ions, base stacking, protein folding, and micelle formation

    B. Hydration of ions, base stacking, and micelle formation

    C. Base stacking and protein folding

    D. Protein folding and micelle formation

    E. Hydration of ions and protein folding

67. **The energy of steric interactions is proportional to**

    A. $1/r$.

    B. $1/r^2$.

    C. $1/r^4$.

    D. $1/r^6$.

    E. $1/r^{12}$.

68. **How is statistical mechanics useful in biophysics?**

    A. Statistical mechanics provides a molecular interpretation of experimental observables.

    B. Statistical mechanics allows us to do statistical analysis of the mechanical forces on a biological system.

    C. Statistical mechanics provides a thermodynamic interpretation of biological events.

    D. All of the above

    E. None of the above

69. **How many distributions of 8 units of energy among four molecules are possible?**

    A. 1

    B. 2

    C. 3

    D. 4

    E. 5

70. **Which distribution shown in Fig. 5-4 is most likely to occur (relative to the other distributions shown in Fig. 5-4)?**

    A. The one with the most permutations

    B. The one with the most molecules

C. The one with the maximum energy available

D. The one with the most distributions

E. None of the above

71. Given the following two expressions for the partition function

$$Z_A = \sum_{j=1}^{NStates} e^{-\frac{E_j}{kT}} \quad \text{and} \quad Z_B = \sum_{i=1}^{L_{MAX}} \omega_i e^{-\frac{E_i}{kT}}$$

which of the following statements is *not* true?

A. In Z_A, each E_j is a different state; In Z_B, each E_i is a different energy level.

B. ω_i is the degeneracy of energy E_i.

C. The value of Z_B will be larger than the value of Z_A because each term is multiplied by ω_i.

D. Whether you use Z_A or Z_B is just a matter of which is mathematically more convenient for the particular statistical mechanical model at hand.

E. If you use Z_B, no two E_i will have the same value, but if you use Z_A, it is possible that some of the E_j will have the same value as other E_j.

72. The *sodium-calcium exchanger,* a membrane transport protein that moves three Na^+ ions into the cell for every Ca^{2+} ion that it moves out of the cell, is an example of

A. inefficient use of sodium.

B. secondary active transport.

C. primary active transport.

D. passive ion channels.

E. the need for calcium to build strong bones.

73. The patch clamp is a technique used in

A. electrophoresis.

B. electrophysiology.

C. X-ray crystallography.

D. optical tweezers.

E. analytical ultracentrifugation.

74. A 225-lb firefighter climbs a ladder straight up to the ninth floor (36 m). Assuming the only energy required to do this is the energy needed to lift the firefighter's weight to the ninth floor, how much energy must the firefighter expend in climbing the ladder?

    A. 36 kcal

    B. 1800 J

    C. 3600 kJ

    D. 36 kJ

    E. 3.6 kJ

75. An increase in entropy

    A. requires energy input to the system.

    B. contributes to an unfavorable Gibbs energy change.

    C. was invented by Rudolf Clausius.

    D. contributes to reducing the Gibbs energy.

    E. explains why living systems can ignore the second law of thermodynamics.

76. Liposomes made from a certain type of phospholipid melt at 310 K. At the melting temperature the Gibbs energy change is zero. The enthalpy change for the lipid phase transition is 62 kcal/mol. What is the entropy change? *Hint:* Use Eq. (4-14).

    A. 200 cal/degree/mol

    B. 2000 cal/degree/mol

    C. 2 kcal/degree/mol

    D. 200 kcal/degree/mol

    E. 22 kcal/degree/mol

77. Liposomes made from a certain type of phospholipid melt at 310 K. At the melting temperature, the Gibbs energy change is zero. The enthalpy change for the lipid phase transition is 31 kcal/mol. Assuming that the enthalpy change and the entropy change are the same at 320 K as they are at 310 K, what is the Gibbs energy change at 320 K? *Hint:* You need to first calculate the entropy change; use Eq. (4-14).

    A. −10 kcal/mol

    B. −20 kcal/mol

C. −1 kcal/mol

D. +1 kcal/mol

E. +10 kcal/mol

78. Which of the following statements is true?

S1. Secondary structure is two dimensional, whereas tertiary structure is three dimensional.

S2. Primary structure is one dimensional, specifying only the sequence in which atoms or groups of atoms are connected to one another.

S3. Biological molecules are always polymers.

S4. Residues are the parts of a molecule left behind after the molecule separates from its parent molecule.

S5. Quaternary structure takes into account time as well as three dimensions in space.

A. S1, S2, S3, and S4

B. S2, S3, S4, and S5

C. S3

D. S2

E. None of the above

79. Enthalpy is internal energy plus

A. pressure times volume.

B. entropy.

C. temperature times entropy.

D. volume times temperature.

E. Gibbs energy.

80. Which of the following is true?

S1. Membrane protein structure is inside out relative to cytoplasmic proteins.

S2. Positive charges on charged amino acid side chains can participate in cation-pi interactions with aromatic amino acid side chains.

S3. Aromatic side chains are most likely to hydrogen bond with water.

S4. A lower potential energy means the conformation is more stable.

  A. S1, S2, and S3

  B. S1, S2, and S4

  C. S2, S3, and S4

  D. S1, S3, and S4

  E. None of the above

81. Eight units of energy are distributed among five molecules as follows: three molecules with 1 unit of energy each; one molecule with 2 units of energy; one molecule with 3 units of energy. How many ways are there to arrange the molecules in the same distribution?

  A. 5

  B. 10

  C. 15

  D. 20

  E. 25

82. If we use the Gibbs energy change to evaluate a biophysical process, we can effectively ignore the second law of thermodynamics because

  A. the second law is irrelevant to biological systems.

  B. the second law only applies when entropy is increasing.

  C. an increase in entropy is favorable for the Gibbs energy change.

  D. the second law is built into the formula for Gibbs energy.

  E. Gibbs was able to show that the second law no longer applies nowadays.

83. What do the two-state approximation and the nearest-neighbor approximation have in common?

  A. They are both observed in DNA.

  B. They are both intermediate steps in calculating the partition function.

  C. Nearest neighbors can only assume two states between them.

  D. All of the above

  E. None of the above

84. Unsaturated lipids are
    A. lipids containing double bonds within the hydrocarbon chain.
    B. lipids that are not saturated with hydrogen.
    C. lipids that cannot pack as closely together as saturated lipids.
    D. all of the above
    E. none of the above

85. A certain receptor on the cell surface binds the hormone insulin with a Gibbs energy change of -12.5 kcal at body temperature (310 K). Using a calorimeter, the binding enthalpy change was measured to be -90 kcal. Calculate the entropy change for insulin binding to the receptor.
    A. -0.1 kcal/degree
    B. -0.75 kcal/degree
    C. -0.25 kcal/degree
    D. -0.025 kcal/degree
    E. -77.5 kcal/degree

86. What is special about pi electrons in aromatic rings?
    A. Each electron contains an internal energy of 3.141593×10^{-20} J.
    B. These elections are called *cations* because they participate in cation-pi interactions.
    C. Pi electrons are delocalized in aromatic rings
    D. They form the circumference of the aromatic ring, which is 2 times π times the radius of the ring.
    E. Nothing (this is a trick question; aromatic rings do not contain pi electrons).

87. Electromagnetic spectroscopy typically measures some property of the EM radiation, such as the amount of radiation that is
    A. scattered.
    B. absorbed.
    C. radioactive.
    D. A and B
    E. B and C

88. Natural DNA sequences are
   A. homogenous.
   B. homozygous.
   C. heterozygous.
   D. heterogeneous.
   E. herbivorous.

89. Which technique can be used to separate molecules both according to size and according to electric charge?
   A. Gel electrophoresis
   B. Atomic force microscopy
   C. Ionization centrifugation
   D. Calorimetry
   E. Mass electrophoresis

90. The peptide bond
   A. links amino acids together.
   B. can be treated as a dipole with $D = 0.7$.
   C. is surrounded on both sides by polar bonds.
   D. all of the above
   E. none of the above

91. In proteins, hydrogen bonds can form directly between various amino acid side chains and other amino acid side chains, or they can form between amino acid side chains and water. The Gibbs energy difference between these two types of hydrogen bonds is small. This fact tells us that
   A. it is easy to form hydrogen bonds in proteins.
   B. amino acid side chains are not very good at forming hydrogen bonds.
   C. many hydrogen bonds are needed to add up to a significant Gibbs energy change.
   D. polar amino acids are an important dietary nutrient.
   E. hydrogen bonding plays a relatively minor role in stabilizing protein structure.

92. **A favorable Gibbs energy change**

    A. means that the Gibbs energy of the system has decreased.

    B. will drive a process forward.

    C. is negative.

    D. all of the above

    E. none of the above

93. **What effect does the presence of cholesterol in a bilayer have on the movement of phospholipids in the plane of the bilayer?**

    A. It causes them to move in a straight line.

    B. It prevents them from moving at all.

    C. It creates steric restrictions on phospholipid movement within the fluid mosaic.

    D. All of the above

    E. None of the above

94. **A change in the shape of a biomolecule is called a**

    A. conformational transition.

    B. topological transition.

    C. phase transition.

    D. transitional configuration.

    E. configurational alignment.

95. **Which has the largest influence on stabilizing protein structure?**

    A. Zwitterions

    B. Hydrophilic dipoles

    C. Hydrophobic forces

    D. Hydrogen bonds

96. **When a *membrane* protein folds into its native state,**

    A. all of the polar amino acid side chains are removed.

    B. nonpolar side chains are on the inside, away from the aqueous environment.

    C. there is a favorable Gibbs energy change for moving the nonpolar side chains to the inside.

    D. hydrophobic portions of the molecule will be on the outside.

    E. the protein will denature rapidly.

97. **When the double helix of a closed duplex DNA unwinds,**
    A. *Tw* increases and *Wr* increases.
    B. *Tw* decreases and *Wr* decreases.
    C. *Tw* increases and *Wr* decreases.
    D. *Tw* decreases and *Wr* increases.
    E. *Tw* and *Wr* remain constant.

98. **Which of the following items apply to eukaryotes?**
    S1. Typically larger
    S2. Typically smaller
    S3. Membrane-bound organelles
    S4. Shorter cell cycle (rapid division)
    S5. Simpler biochemical reactions
    A. S1 and S3
    B. S2 and S4
    C. S1 and S4
    D. S2 and S3
    E. S2, S4, and S5

99. **If a dipole is allowed to rotate freely in solution, what technique is used to calculate the interaction energy?**
    A. Circular dichroism
    B. Rotational symmetry
    C. Angular averaging
    D. Thermal averaging
    E. Rotational averaging

100. **What does the cytoskeleton do?**
    A. Supports skin and muscle cells
    B. Adds support to help the cell maintain its shape
    C. Is involved in transporting molecules from one location in the cell to another
    D. A and B
    E. B and C

Answers to Quizzes and Final Exam

Chapter 1	Chapter 3	Chapter 5	Chapter 7
1. C	1. C	1. D	1. B
2. D	2. A	2. B	2. C
3. A	3. C	3. A	3. D
4. D	4. C	4. D	4. B
5. D	5. C	5. B	5. D
6. C	6. C	6. A	6. A
7. D	7. B	7. D	7. A
8. A	8. D	8. D	8. B
	9. C	9. A	9. D
	10. B	10. C	10. D

Chapter 2	Chapter 4	Chapter 6	Chapter 8
1. C	1. C	1. C	1. D
2. D	2. A	2. C	2. A
3. B	3. D	3. E	3. D
4. D	4. C	4. B	4. B
5. C	5. B	5. D	5. B
6. A	6. D	6. D	6. C
7. D	7. D	7. C	7. D
8. D	8. D	8. A	8. C
9. D	9. B	9. E	9. C
10. C	10. A	10. A	10. G

Chapter 9
1. A
2. C
3. C
4. C
5. C
6. A
7. A
8. D
9. B
10. D

Chapter 10
1. C
2. D
3. A
4. D
5. A
6. B
7. C
8. D
9. F
10. C

Chapter 11
1. B
2. A
3. D
4. A
5. B
6. D
7. B
8. D
9. A
10. C

Chapter 12
1. A
2. D

3. B
4. B
5. A
6. C
7. D
8. C
9. D
10. A

Final Exam
1. C
2. A
3. A
4. E
5. B
6. C
7. C
8. D
9. D
10. E
11. C
12. E
13. D
14. D
15. D
16. C
17. B
18. A
19. E
20. A
21. A
22. A
23. D
24. B
25. D
26. D
27. A
28. D

29. C
30. E
31. C
32. E
33. B
34. D
35. C
36. D
37. D
38. A
39. D
40. C
41. A
42. C
43. B
44. C
45. E
46. A
47. D
48. C
49. D
50. C
51. C
52. C
53. D
54. D
55. B
56. A
57. B
58. C
59. E
60. E
61. D
62. C
63. B
64. D
65. C
66. D

67. E
68. A
69. E
70. A
71. C
72. B
73. B
74. D
75. D
76. A
77. C
78. D
79. A
80. B
81. D
82. D
83. E
84. D
85. C
86. C
87. D
88. D
89. A
90. D
91. E
92. D
93. C
94. A
95. C
96. D
97. D
98. A
99. D
100. E

appendix

Units, Conversions, and Constants

Common Prefixes

Prefix	Symbol	Factor
Femto	f	10^{-15}
Pico	p	10^{-12}
Nano	n	10^{-9}
Micro	μ	10^{-6}
Milli	m	10^{-3}
Centi	c	10^{-2}
Kilo	k	10^{3}
Mega	M	10^{6}
Giga	G	10^{9}
Tera	T	10^{12}

Standard Units

Quantity	Unit	Symbol	Equivalent to
Length	Meter	m	
Velocity, speed	Meter per second	m/s	
Acceleration	Meter per second squared	m/s^2	Meter per second per second
Mass	Kilogram	kg	
Time	Second	s	
Temperature	Kelvin	K	
Amount	Mole	mol	6.022×10^{23}
Current	Ampere	A	
Force	Newton	N	$kg \cdot m/s^2$
Pressure	Pascal	Pa	$N/m^2 = m^{-1} \cdot kg \cdot s^{-2}$
Energy, work	Joule	J	$N \cdot m = kg \cdot m^2/s^2$
Power	Watt	W	J/s
Current	Ampere	A	C/s
Electric charge	Coulomb	C	$A \cdot s$
Electromotive force (electrical potential)	Volt	V	J/C
Electrical resistance	Ohm	Ω	V/A
Volume	Liter	L	10^{-3} m^3
Concentration	Molarity	M	mol/L

Some Conversions for Nonstandard Units

Quantity	Standard Unit	Nonstandard Unit	Conversion
Energy	Joule	Calorie	1 cal = 4.184 J
Force	Newton	Pound	1 lb = 4.4482 N
Length	Meter	Inch	1 in = 0.0254 m
Length	Meter	Angstrom	1 Å = 10^{-10} m
Pressure	Pascal	Torr (mm Hg)	1 torr = 133.322 Pa
Temperature	Kelvin	Centigrade	°C = K − 273.15
Volume	Meter^3	Liter	1 L = 10^{-3} m^3

Some Useful Constants

Constant	Symbol	Value
Boltzmann's constant	k_B	1.3806×10^{-23} J/K
Planck's constant	h	6.626068×10^{-34} J·s
Coulomb's constant	k_e	8.98755×10^9 N·m²/ C²
Speed of light	c	2.9979×10^8 m/s
Avagadro's number	mol	6.022×10^{23}
Acceleration due to gravity	g	-9.807 m/s²

Some Useful Formulas

Formula	Meaning	Equation
$F = ma$	Force = mass times acceleration, Newton's second law of motion	1–1
$E = h\upsilon$	Energy of electromagnetic radiation is equal to Planck's constant times the frequency of the radiation	3–2
$H = U + PV$	Enthalpy = internal energy + pressure times volume	4–3
$\Delta S = Q/T$	Entropy change = heat flow divided by temperature	4–11
$\Delta G = \Delta H - T\Delta S$	Gibbs energy change	4–14
$W = \dfrac{N!}{n_1! \cdot n_2! \cdot n_3! \cdots n_{L_{MAX}}!}$	Number of ways to arrange N objects into L_{MAX} groups with n_i items in group i	5–1a
$W = \dfrac{N!}{\prod\limits_{i=1}^{L_{MAX}} (n_i!)}$	Number of ways to arrange N objects into L_{MAX} groups with n_i items in group i	5–1b
$Z = \sum\limits_{i=1}^{L_{MAX}} e^{-\beta E_i}$	Partition function	5–4
$F_i = \dfrac{n_i}{N} = \dfrac{e^{-\frac{E_i}{kT}}}{Z}$	Fraction of molecules in state i	5–6

(Continued)

Formula	Meaning	Equation
$\langle E \rangle = \sum_{i=1}^{L_{MAX}} E_i \cdot F_i = \sum E_i \cdot \dfrac{e^{-\frac{E_i}{kT}}}{Z}$	Average energy	5–8
$F = k_e \dfrac{q_1 q_2}{r^2}$	Force between two electrostatic charges	6–1
$U = \dfrac{q_1 q_2}{\varepsilon r}$	Potential energy between two electrostatic charges	6–2b
$\alpha_p = \dfrac{d}{E}$	Polarizability	6–10
$Z = 1 + e^{-\frac{\Delta H}{RT} + \frac{\Delta S}{R}}$	Partition function for a two-state transition	10–2
$\sqrt{\langle r^2 \rangle} = \sqrt{N} \cdot L$	RMS (root-mean-square) end-to-end distance in a freely jointed chain	10–4
$Wr = Lk - Tw$	Writhe = linking number − twist	10–5
$a = (v_2 - v_1)/t$	Acceleration	12–2
$v_2^2 = v_1^2 + 2\,a\,d$	Final velocity in terms of initial velocity, acceleration, and distance traveled	12–9
$P + \rho\,gh + (1/2)\rho\,v^2 = $ constant	Bernoulli's equation for a frictionless fluid	12–12
$d = (1/2)\,a\,t^2$	Distance traveled under a constant acceleration if the initial velocity is zero	12–24

Glossary

Absorbance A measure of how much light does *not* pass through a sample. Commonly reported in terms of molar *extinction coefficient*, or how much light is absorbed per concentration per length of sample that the light passes through.

Absorption spectroscopy A form of *EM spectroscopy* in which we shine light of known wavelengths through a sample and measure the intensity of the light that comes out the other side.

Acid A molecule or functional group that increases the concentration of hydrogen ions (protons) in solution.

Active site The part of a molecule or *complex* that is involved in carrying out its function. A molecule or complex can sometimes have more than one active site, depending on the function of the molecule.

Active transport Transport in which the cell expends energy to move a molecule or ion across a membrane. See also *primary active transport* and *secondary active transport*.

Adiabatic system A *closed system* that is thermally isolated from its *surroundings*; there is no exchange of heat or matter, but it is possible for the system to exchange nonheat energy with its surroundings. For example, the system can do work on its surroundings (or have work done on it).

Aldehyde A carbonyl carbon that is also directly covalently bonded to at least one hydrogen atom.

Allosteric protein A protein that exhibits the property of *allostery*.

Allosteric regulation Regulation of a biochemical process via the allosteric site on a molecule.

Allosteric site The part of a molecule or complex, other than the active site, that influences the *active site* of the molecule.

Allostery A property of a molecule, usually of a protein, whereby the biological activity of one part of the molecule (the *active site*) is affected by some action or intermolecular binding at another part of the molecule (the *allosteric site*). The ability of the molecule's *active site* to carry out its function is dependent on binding or activity at some other site (the *allosteric site*) on the molecule.

Alpha carbon In an amino acid, the carbon atom that has the amine group and carboxyl group attached to it.

Alpha helix A common secondary structure in proteins in which the polypeptide chain forms a right-handed helical shape. The backbone of the chain is on the inside of the helix with the various amino acid side chains on the outside.

Amine group A nitrogen atom bonded in such a way as to have a pair of nonbonded valence electrons. An *amine* nitrogen atom has three single covalent bonds to three other atoms.

Amino acid Any of a class of biochemical compounds that contain both an amine group and a carboxylic acid group (carboxyl group) attached to a single carbon atom. This carbon atom (called the *alpha carbon*) also has a *side chain* (some other group of atoms) attached to it. There are 20 or so different amino acids found in living things, and each is named according to its side chain.

Amphipathic Having both hydrophobic and hydrophilic properties in the same molecule; sometimes also called *amphiphilic*.

Amphiphilic Same as amphipathic

Analytical techniques A technique used to measure physical aspects of a biological system.

Analytical ultracentrifuge An *ultracentrifuge* that contains special optical devices and sensors that can track the movement of molecules as they are being centrifuged.

Anion A negative ion; an atom or molecule that has negative charge due to having one or more extra electrons.

Antibody Also called *immunoglobulin,* any of a class of proteins that can recognize and bind to foreign objects (i.e. antigens, such as viruses and bacteria).

Antiport *Cotransport* where at least some of the different types of transported molecules or ions are being transported in opposite directions across a biological membrane.

Aromatic A molecular ring structure having (1) all of the atoms making up the ring lying within a plane and (2) all of the ring atoms contributing electrons from their *pi orbital* to make up the covalent bonds that hold the ring together. In an aromatic structure, the *pi electrons* (electrons from the *pi orbital*) are delocalized; that is, they are shared among *all* of the atoms in the ring.

Atomic force microscopy (AFM) A technique to provide highly magnified images similar to those of SEM (i.e., images of the surface of a specimen) but with resolution similar to that of TEM (about 10 times better than SEM).

ATPase Any enzyme that couples a process or reaction to hydrolysis of ATP and utilizes the energy from the hydrolysis of ATP to drive the process or reaction.

Available energy Energy that can be transformed into work, where work equals force times distance. *Available energy* is also called *free energy*, and *Gibbs energy*.

Base pairing Hydrogen bonding between nucleotide bases in nucleic acids such that adenine and thymine (or adenine and uracil, in the case of ribonucleic acids) always pair together, and guanine and cytosine always pair together.

Base stacking *Stacking interaction* among nucleotide bases.

Base A molecule or functional group that decreases the concentration of hydrogen ions (protons) in solution. Bases often decrease the concentration of hydrogen ions by increasing the concentration of hydroxyl (OH^-) ions in solution. The hydroxyl ions then combine with the hydrogen ions to form water.

Beta sheet A common secondary structure in proteins in which two polypeptide strands are hydrogen-bonded to one another. Also called *beta pleated sheet*.

Binding site A particular location on a molecule that binds to another molecule or ligand.

Biochemical physics See *Molecular and Subcellular Biophysics*.

Biophysical chemistry See *Molecular and Subcellular Biophysics*.

Biomechanics The branch of biophysics that deals with the application of forces to biological objects.

Biopolymer A biomolecule that is a polymer.

Boltzmann distribution A particular distribution derived by Ludwig Boltzmann and containing the maximum the number of permutations.

Budding Extrusion, or pushing out, of a portion of a membrane until it pinches off to form a new liposome.

Calorimeter An instrument designed to measure heat energy and other thermodynamic properties.

Calorimetry Any of a series of techniques using instruments designed to directly measure thermodynamic properties such as heat.

Carbohydrate A molecule made of carbon, hydrogen, and oxygen, in the ratio $C_nH_{2n}O_n$. See also *dexyocarbohydrate*.

Carbonyl group A carbon atom with a double covalent bond to an oxygen atom.

Carboxyl group A carbonyl carbon that is also covalently connected to a hydroxyl group.

Carboxylic acid A substance whose molecules contain a carboxyl functional group.

Catalyst A substance that modifies the rate of a chemical reaction, without actually being consumed by the reaction.

Cation A positive ion; an atom or molecule that is missing one or more electrons.

Cation-pi interaction An attractive interaction between a *cation* (positive ion) and the partial negative change of the *pi electrons* of an aromatic molecule.

Cell A pouch or sack of biomolecules where the entire pouch is (or was) alive.

Cell cycle The phases of cell growth and cell division. The two main phases of the cell cycle are *interphase* and *cell division*. A third phase, called *quiescent*, means that the cell has essentially left the cycle and has stopped growing and dividing, but continues to utilize energy to do work.

Cell division The process whereby cells come from preexisting cells, by replicating their DNA and then dividing in two.

Cell theory The two tenets of cell theory are: (1) All living things are made of one or more cells, and (2) all cells come from preexisting cells.

Cell wall A relatively stiff protective layer on the outside surface of the cell membrane of some cells. Commonly found on plant cells and many prokaryotic cells, but rarely on animal cells.

Centrifuge A machine used to spin a sample of material around in a circular motion in order to apply force to the sample.

Chaperone A protein that acts as a *folding moderator.*

Chlorophyll A class of molecules that absorb light for photosynthesis.

Chloroplasts Organelles that carry out *photosynthesis;* found in plant cells and a few other eukaryotes.

CHNOPS An acronym for the elements most commonly found in living cells: carbon, hydrogen, nitrogen, oxygen, phosphorus, and sulfur.

Chromatin All of the DNA molecules in a cell together, in their tertiary or quaternary structures, but not in their condensed state appearing as individual chromosomes (as they are sometimes), but rather in a state where they are more spread out, like a plate of spaghetti, and it is difficult to detect one DNA molecule from another.

Chromosome A DNA molecule as it exists in the cell in its organized and compact tertiary or quaternary structure (which may also include associated proteins).

Circular DNA A DNA molecule in which then ends of the DNA strand are covalently bonded together, forming a circle.

Closed duplex DNA (cdDNA) A circular, double-stranded, double-helical DNA in which the two ends of the double helix are covalently bonded together forming a double-helical circle,

Closed system A *system* that can exchange only energy (not matter) with its *surroundings.*

Command voltage The value of the voltage that is held constant by a voltage clamp.

Complementary A property of two nucleic acid sequences such that the sequence of nucleotides in one nucleic acid sequence can perfectly *base-pair* with the nucleotides in another nucleic acid sequence.

Complex A quaternary structure. The term *complex* usually implies that the at least two different molecules make up the complex; however, the term may sometimes be used for a quaternary structure in which all of the component molecules are identical.

Conformational transition A change in the shape (conformation) of a molecule.

Constructive interference The interference of two or more waves, specifically in a way that results in a larger or more intense wave (the peaks become higher and the troughs become lower).

Cooperative unit length In a polymer undergoing a *cooperative* transition, the number of residues that tend, on average, to undergo the transition together.

Cooperativity The occurrence of separate events together in a nonindependent, relatively all-or-none manner.

Cotransport Active transport that involves transporting more than one type of molecule or ion.

Cotransporter A transport protein that mediates the simultaneous transport (*cotransport*) of more than one type of molecule or ion.

Covalent bond A bond that holds two atoms together by virtue of one or more pairs of valence electrons being shared by the two atoms.

Cristae Membranous folds that fold in toward the center of a *mitochondrion*.

Critical flow velocity The velocity at which fluid flow changes from laminar to turbulent.

Cross-linking A covalent connecting of otherwise distant parts of a molecule.

Crystal An orderly, three-dimensional, repeating arrangement of atoms or molecules.

Cumulative probability The combined probability (the sum of the probabilities) of one or more states or distributions of a system, from the first state or distribution, up to and including a specified state or distribution of the system. For cumulative probability to be meaningful, the sequence of the states or distributions (i.e., the definition of first, last, and how to get from first to last) must be specified.

Current clamp A technique used in electrophysiology to measure and characterize electric voltages across cell membranes by holding the current constant.

Cytoplasm The inside of a cell (excluding the inside of membrane-bound organelles; although membrane-bound organelles, such as the nucleus and mitochondria, are themselves in the cytoplasm, the insides of membrane-bound organelles are typically not considered part of the cytoplasm).

Cytoskeleton Fibrous tubelike and ropelike, sometimes interconnected, structures inside cells that support, transmit, or apply forces within the cell. (See also *microtubule*, *intermediate filament*, *microfilament*, and *tubulin*.)

Degeneracy In statistical mechanics, for a particular energy level, the number of different states of a system that have that same particular energy.

Delocalized electrons Electrons that are able to move about an entire molecule, or section of a molecule (as opposed to being held within the region surrounding only one atom, or two adjacent covalently bonded atoms).

Denatured The state of a protein or nucleic acid that has unfolded, so that it is no longer in its native state.

Denaturing agent A substance or environmental condition that destabilizes a higher-order structure (quaternary, tertiary, and/or secondary structure) in proteins or nucleic acids, causing them to unfold out of their native state.

Deoxycarbohydrate A carbohydrate that is missing one oxygen atom, that is, a molecule made of carbon, hydrogen, and oxygen, in the ratio $C_nH_{2n}O_{n-1}$. See also *carbohydrate*.

Deoxyribose A 5-carbon deoxycarbohydrate, whose ring form is commonly found in nucleotides.

Destructive interference The interference of two or more waves specifically in a way that results in a smaller or less intense wave (the distance between the peaks and troughs becomes smaller).

Dielectric constant A measure of how easily a medium (material or substance) is polarized by an electric field. The dielectric constant therefore also measures how easily a material or substance conducts electricity. Dielectric constant is also called *permittivity*.

Differential scanning calorimeter An instrument that can measure the heat capacity of a sample as a function of temperature.

Differentiation The process of a cell specializing and becoming a particular type of cell.

Diffraction A phenomenon whereby light waves pass through an ordered arrangement of openings and subsequently interfere with each other on the other side.

Diffusion The process of molecules or particles spreading out as a result of their random motion. In diffusion, molecules or particles always move (over time) from a region of higher concentration to one of lower concentration, however, not necessarily directly or straight away, but rather somewhat gradually as the result of random motion.

Dipole moment The magnitude of the charge at either end of an electric dipole times the fixed distance between the two charges at the opposite ends of the electric dipole.

Disaccharide Two monosaccharides covalently bonded together into a single molecule.

Disorder In thermodynamics and statistical mechanics, order and disorder relate directly to the *number of different ways of accomplishing one particular thing*. An ordered system is one in which there is a small number of ways to accomplish a particular state of the system. A disordered system is one in which there are many ways to accomplish a particular state of the system.

Dispersion forces Intermolecular forces that arise from synchronized fluctuating charge distributions.

Distribution In *statistical mechanics*, a specification of how the total energy of a system is divided up among the particles or molecules of the system, in terms of how many molecules have specified amounts of energy.

Disulfide bond A covalent bond between two sulfur atoms, attaching two cysteine residues together.

DNA (deoxyribonucleic acid) A nucleic acid whose nucleotides contain deoxyribose.

DNA melting Unwinding or denaturing of the DNA helix.

DNA polymerase An enzyme that catalyzes the biochemical reaction of covalently connecting together deoxyribonucleotides to form DNA.

DNA replication The processes whereby a cell makes a copy of its DNA.

Double bond A covalent bond in which two pairs of electrons are shared between two atoms.

Double helix The most common secondary structure in DNA, consisting of two helical strands wound around each other.

Eddies Small fluid currents that flow backward, or in a partial circular motion, opposite the direction of the overall fluid flow.

Electric dipole Two equal and opposite charges separated by a fixed distance.

Electrolyte solution A solution containing a relatively high concentration of ions, so that it easily conducts electricity.

Electronegative Having the quality of a strong pull on electron clouds. Electronegative atoms have a tendency to form polar bonds when covalently bonded to less electronegative atoms.

EM spectroscopy Electromagnetic spectroscopy. A technique that involves sending some form of electromagnetic radiation into a sample, and measuring various properties of the electromagnetic radiation that emerges from the sample. The measured property is then plotted as a function of the wavelength or frequency of the radiation; the resulting plot is called a *spectrum* (plural: *spectra*). See also *spectroscopy*.

Endocytosis A process whereby a portion of the cell membrane folds in toward the cell, engulfing some molecule or molecules from outside the cell. This portion of membrane then breaks off from the membrane, forming a *vesicle* that carries the engulfed contents into the cell.

Endomembrane system A set of membranous organelles that work together in eukaryotic cells; the endomembrane system includes the endoplasmic reticulum, Golgi apparatus, nuclear envelope, and other membranous organelles.

Endoplasmic reticulum A network of folded membranes, in eukaryotic cells, that provides a surface on which various biochemical reactions are carried out.

Energy path How the energy of a system changes (ups and downs) over the course of time or throughout the duration of a process.

Enthalpy A thermodynamic quantity equal to internal energy plus pressure-volume energy.

Enthalpy driven A spontaneous (favorable) process in which the change in enthalpy makes a favorable contribution to the Gibbs energy change, but the entropy change is either unfavorable or negligible. The enthalpy change is primarily responsible for the favorable Gibbs energy change.

Entropy The amount of energy contained in a system that cannot be transformed into work. Entropy is also a measure or the amount of disorder in a system: the more disordered the energy of a system is, the less of that energy can be transformed into work.

Entropy driven A spontaneous (favorable) process in which the change in entropy makes a favorable contribution to the Gibbs energy change, but the enthalpy change is either unfavorable or negligible. The entropy change is primarily responsible for the favorable Gibbs energy change.

Environmental biophysics The branch of biophysics that deals with physics of the relationship between organisms and their environment.

Enzyme A protein that acts as a catalyst.

Ester A type of carbonyl group very similar to carboxyl, but where the hydroxyl hydrogen or the carboxyl group has been replaced with a carbon atom.

Eukaryote An organism that has *eukaryotic cells.*

Eukaryotic The quality of having a cell nucleus.

Excitable tissue Cell types, in an organism, that create or conduct electric impulses.

Exocytosis A process whereby a vesicle inside the cell fuses with the plasma membrane, releasing its contents to the environment outside the cell.

Expression Transcription of a gene's sequence into mRNA, followed by translation of the mRNA into the polypeptide or protein.

Extinction coefficient (or *molar extinction coefficient*) A measure of absorbance that accounts for both the concentration and the thickness of the sample being studied; the absorbance is expressed in units that are per concentration and per length ($M^{-1}cm^{-1}$).

Fatty acid A long chain hydrocarbon with a carboxyl (carboxylic acid) group at one end. (See also *saturated, unsaturated,* and *polyunsaturated fatty acid.*)

Favorable A process is favorable, or spontaneous, if the Gibbs energy change for that process is negative; that is, if the process decreases the Gibbs energy.

First law of thermodynamics Any change to the amount of energy contained in a system is equal to the amount of energy put into the system minus the amount of energy taken out of the system. In other words, energy cannot be created or destroyed.

Fluid mosaic model A model that describes biological membranes as composed of a phospholipid bilayer with other molecules scattered about within the plane of the bilayer.

Fluorescence The emission of light, or electromagnetic radiation, from a molecule or substance.

Fluorescent tagging A technique in which a fluorophore is attached to another molecule in order to follow that molecule through some biological process.

Fluorophore A molecule or the specific part of a molecule that is responsible for the fluorescence.

Folding moderators Molecules that assist in protein folding. Also called *chaperones.*

Free energy Energy that can be transformed into work, where work equals force times distance. Also called *available energy* and *Gibbs energy.*

Freely jointed chain A model in which each segment or residue of a chain is free to angle and rotates relative to its adjacent segments or residues.

Fuel A source of potential energy for a *motor*.

Functional group A specific group of atoms within a molecule that confers certain functional characteristics to that molecule.

Gel A fluid that has a molecular structure that gives it properties similar to a solid.

Gel electrophoresis A technique used to *sediment* molecules through a *gel* via the application of an electric field.

Gel phase The relatively solid state of a lipid bilayer.

Gene A section of a DNA molecule that specifies the amino acid sequence for a given protein.

Genome All of a cell's genes (on all of its chromosomes or chromatin) taken together as a whole.

Gibbs energy Energy that can be transformed into work, where work equals force times distance. Also called *available energy* and *free energy.*

Gibbs energy change The enthalpy change minus the quantity of the absolute temperature times the entropy change $\Delta G = \Delta H - T\Delta S$ where ΔH is the enthalpy change, T is the absolute temperature, and ΔS is the entropy change.

Glyceride Any of a class of molecules consisting of a glycerol molecule with a fatty acid attached to one or more of glycerol's three carbon atoms. The result is a monoglyceride, diglyceride, or triglyceride.

Glycerol A small, three-carbon carbohydrate, with a hydroxyl group attached to each carbon atom.

Ground state The lowest energy state of a system.

Gyrase A type of topoisomerase that catalyzes an increased negative writhe by reducing the linking number.

Heat capacity The amount of heat required to raise the temperature of a sample by one degree.

Heat energy Energy that results from the kinetic energy and random motion of molecules.

Heat engine A *motor* that converts *heat energy* into motion.

Helical repeat In a helical polymer, the number of residues in a single turn of the helix.

Helicase An enzyme that catalyzes the unwinding of DNA.

Helix-coil transition A conformational transition from a helix conformation to a random-coil conformation; unwinding of the helix.

Heterogeneous sequence A sequence that varies throughout the molecule and is not repetitive.

Heteropolymer A polymer in which the residues are *not* all identical.

High-energy intermediate If a process causes a system to go through intermediate states along the way from its initial state to its final state and if one or more intermediate states are higher in energy than the initial and end states, then those intermediate states that are higher in energy than the initial and end state are called *high-energy intermediates*.

Histone A cluster (quaternary structure) of certain proteins that are involved in DNA packaging by having the DNA bound to and wrapped around them.

Homogenous sequence A simple and repetitive sequence throughout the molecule.

Homopolymer A polymer in which the residues are all identical.

Hydrocarbon A molecule made up of only hydrogen and carbon.

Hydrocarbon chain A hydrocarbon molecule in which the carbon atoms are covalently connected to one another in a line, like links in a chain.

Hydrogen bond The attractive force between a hydrogen atom with a partial positive charge and another atom with a partial negative charge. The two atoms participating in the hydrogen bond (i.e. the hydrogen and the other atom) are not directly covalently bonded together, nor are they ever covalently bonded with only a single atom between them.

Hydrophilic Having polar properties that allow a molecule to interact favorably (i.e., with a negative Gibbs energy change) with water, for example, forming hydrogen bonds with the water.

Hydrophilic force A close-range repulsive force keeping the two hydrophilic surfaces apart, due to a layer of water molecules between them, that results from their strong tendency to hydrogen-bond with water.

Hydrophobic Having nonpolar properties, and thus interacting unfavorably (i.e., with a positive Gibbs energy change) with water.

Hydrophobic effect or hydrophobic interaction An interaction between water and a nonpolar substance like oil that results in configurations that minimize contact between the water and the nonpolar substance.

Hydroxyl group An oxygen atom with one hydrogen atom covalently attached.

Immunoglobulins Also called *antibodies*, proteins that can recognize and bind to foreign objects (such as viruses and bacteria).

Induced dipole A distortion of the electron cloud of a neutral, nonpolar molecule due to the presence of an electric field, so that the molecule becomes a dipole.

Interference The combining of two or more waves in a specific location resulting in increased or decreased wave height at that location. See also *constructive* and *destructive interference*.

Intermediate filament A mid-sized tubelike or ropelike structure that is part of the cytoskeleton.

Interphase The phase of the *cell cycle* where most of the cell's energy goes into growth and DNA replication.

Ion An atom or molecule with a charge, due to one or more missing electrons or due to one or more extra electrons.

Ion channel A type of transport protein that spans the bilayer and folds in a way to create a hydrophilic tunnel through which ions and other small hydrophilic molecules can pass.

Ionic bond A bond that holds two atoms together due to each atom being an ion, one positive (cation) and one negative (anion).

Ketone A carbonyl carbon that is also directly connected to two other atoms neither of which is hydrogen.

Laminar flow Fluid flow in which the velocity of the fluid varies in layers, or according to the distance from the vessel walls, due to friction.

Left-handed helix A helix that turns counterclockwise as you move along the length of the helix.

Ligand A smaller molecule or atom that binds to a larger molecule.

Ligand binding An association of one or more ligands (smaller molecules) with a larger molecule, held together by non-covalent forces.

Linking number *(Lk)* The number of times that one closed curve is linked to another closed curve.

Lipid Fat or oil. A molecule characterized by being not soluble in water and soluble in nonpolar solvents.

Lipid vesicle A small hollow sphere of lipid molecules. The lipid molecules making up the vesicle are organized into a *lipid bilayer.*

Lipoprotein A complex of protein and lipid, used to carry lipids to where they are needed and able to carry lipids through places where the lipids would otherwise be insoluble.

Liposome A lipid bilayer sphere with a void in the middle, also called a *vesicle or lipid vesicle*.

Liquid crystalline phase The relatively fluid phase of a lipid bilayer.

Lysosome A vesicle that contains enzymes (called *lysozymes*) that specialize in breaking down or digesting larger molecules.

Machine A device that can alter the direction and/or size of a force

Macromolecule A very large molecule, often a polymer such as DNA, proteins, carbohydrates, and phospholipids. Although there is no clear cutoff in size that defines a molecule as a macromolecule, common usage of the word is typically reserved for molecules that contain more than 1000 atoms.

Mechanics The branch of physics that deals with the application of forces to physical objects.

Membrane A thin flexible sheetlike structure that separates things. In the case of a biological cell, membranes separate between the cell and the outside world or within the cell between different regions of the cell.

Membrane protein A protein that has formed a quaternary structure with a membrane so as to become part of the membrane.

Membrane transport protein A membrane protein that is involved (passively or actively) in transporting other molecules across the membrane.

Micelle A lipid ball with the hydrocarbon chains pointed in toward the center, and the polar head groups at the surface of the ball.

Microfilament The smallest of several tubelike or ropelike structures that are part of the cytoskeleton.

Microtubule The largest of several tubelike structures that are part of the cytoskeleton.

Mitochondrion (plural: *mitochondria*) An organelle that carries out the role of generating ATP molecules for use as energy currency throughout the cell.

Model A hypothesis, guess, or theory, with some details about the mechanism of how a system undergoes a certain process.

Molar extinction coefficient See *extinction coefficient*.

Molecular and subcellular biophysics The branch of biophysics that deals with the behavior of biomolecules and the workings of individual cells.

Molecular binding An association of two or more molecules, or subunits, held together by non-covalent forces.

Molecular interpretation An explanation of what the molecules or particles of a system are doing, when the system undergoes some overall change.

Monosaccharide A single unit carbohydrate (not a polymer), also called a *simple sugar*. (See also *disaccharide, oligosaccharide,* and *polysaccharide*.)

Most probable distribution In statistical mechanics, the distribution that has the largest number of ways to arrange the molecules, that is, the largest number of permutations.

Motor A *machine* that also has the ability to convert potential energy into mechanical energy (motion).

mRNA (messenger RNA) A single-stranded RNA molecule that contains a complementary copy of the sequence of nucleotides in a particular section of DNA.

Native state The state of a protein or nucleic acid that has folded into its correct functional shape where its Gibbs energy has been minimized. The native state can include one or a small number of conformations.

Nearest-neighbor approximation An approximation that assumes only adjacent residues can influence each other, but residues at a distance (or nonadjacent) cannot.

Negatively supercoiled A superhelix in which *writhe* is negative.

Nicked duplex DNA (ndDNA) A circular, double-helical DNA with one or more breaks in one or more strands.

Nitrogenous base A class of organic base compounds containing nitrogen and owing their property as a base (a compound that reduces H^+ ions) to a pair of nonbonded electrons on a nitrogen atom. Nucleic acids contain nitrogenous bases with aromatic ring structures categorized as *purines* and *pyrimidines*.

Nonpolymeric lipid A lipid that is not a polymer or that does not contain a polymer as a major portion of its structure. A *nonpolymeric* lipid is also called a nonsaponifiable lipid, meaning that it cannot be broken down by the chemical process of hydrolysis.

Nuclear envelope A structure that includes both the inner and outer membranes of the nucleus and the *perinuclear space*.

Nuclear lamina A network of intermediate filaments, just inside the inner nuclear membrane, that plays a role similar to the cytoskeleton.

Nuclear pore A membrane transport protein that spans the nuclear envelope and controls the flow of molecules into and out of the nucleus.

Nucleic acid A polymer of *nucleotides*, so named for being an *acidic* substance originally discovered in the *nucleus* of cells.

Nucleic acid palindrome A nucleotide sequence that is its own complement when read backward.

Nucleosome A DNA double helix wrapped around a *histone*.

Nucleotide Any of a class of molecules with (1) a *nitrogenous base* (also called a *nucleotide base*), (2) a sugar (*ribose* or *deoxyribose*)*,* and (3) one to three phosphate groups.

Nucleotide base Any of the nitrogenous bases found in nucleic acids. A *purine* or *pyrimidine*.

Nucleus A central, membrane-bound compartment, in eukaryotic cells, which contains most or all of the cell's genetic material.

Oligosaccharide A carbohydrate polymer containing between 2 and 20 monosaccharides. (See also *polysaccharide*.)

Open system A system that can exchange both matter and energy with its *surroundings*.

Optical trap The phenomenon of focused laser beams holding a single particle in place in three dimensions.

Optical tweezers An instrument that uses focused laser beams to create piconewton (10^{-12} N)-size forces used to hold and manipulate microscopic particles (even as small as a single molecule or atom).

Orbital A mathematically defined region in three-dimensional space that describes where a particular electron, or set of electrons, is most likely to be found orbiting the nucleus of an atom.

Order In thermodynamics and statistical mechanics, order and disorder relate directly to the *number of different ways of accomplishing one particular thing*. An ordered system is one in which there is a small number of ways to accomplish a particular state of the system. A disordered system is one in which there are many ways to accomplish a particular state of the system.

Partition function A mathematical function used in statistical mechanics that is the sum of the statistical weights of each of the energy states of the system and that describes mathematically how the total energy is partitioned among the particles of the system.

Passive transport Transport of a molecule or ion through the membrane by virtue of a favorable Gibbs energy change without the need for the cell to expend any energy.

Patch clamp A technique for applying an electrode to a cell, in which the electrode is placed inside a micropipette (a very thin glass tube) filled with an *electrolyte solution* and the micropipette is placed against the cell membrane.

Peptide bond The covalent bond that links together amino acids in a protein. The peptide bond forms between the carboxyl carbon of one amino acid and the amine nitrogen of the next amino acid.

Perinuclear space The space between the inner nuclear membrane and the outer nuclear membrane. (The nucleus has two membranes; see also *nuclear envelope*.)

Peripheral membrane protein A *membrane protein* that is, for the most part, just on the surface of the membrane (but is still fundamentally part of the membrane).

Permittivity A measure of how easily a medium (material or substance) is polarized by an electric field. Permittivity therefore also measures how easily a material or substance conducts electricity. Permittivity is also called *dielectric constant*.

Permutation Any of a number of equivalent rearrangements of a set.

Peroxisome A vesicle that contains enzymes that specialize in breaking down long-chain fatty acids.

Persistence length The length of a segment of polymer that behaves as if it is relatively rigid.

Phosphate group A phosphorous atom with four oxygen atoms covalently attached.

Phospholipid A class of lipid molecule consisting of phosphate group attached to a *glyceride*.

Photosynthesis The conversion of light energy into chemical bond energy.

Physical biochemistry See *molecular and subcellular biophysics*.

Pi bond A covalent bond in which two orbitals, each with two lobes, overlap both lobes.

Pi orbital A type of orbital with a shape that has two lobes.

Pitch The length of a single turn of a helix.

Plasma membrane The cell membrane.

Polar bond A covalent bond in which one or more electrons are shared unevenly, so that the electric charge is not evenly distributed across the bond. The result is that one atom has a partial positive charge and one atom has a partial negative charge.

Polarizability For an induced dipole, the ratio of the dipole moment to the electric field strength.

Polymer A large molecule made by connecting together many smaller molecules. See also *residue*.

Polypeptide A polymer of amino acids; a protein.

Polysaccharide A carbohydrate polymer containing more than 20 *monosaccharides*. (See also *oligosaccharide*.)

Polyunsaturated fatty acid A *fatty* acid with two or more double bonds in the hydrocarbon chain.

Preparative technique A technique used to purify or isolate biological specimens (organisms, cells, or molecules) or otherwise get them ready for use in some other process or further experimentation.

Primary active transport An *active transport* where the energy needed for transport comes from a simultaneous reaction cleaving a high-energy phosphate bond.

Primary amine An amine where two of the three covalent bonds are to hydrogen, while the third atom is to a carbon atom, linking the amine group to the rest of the molecule.

Primary structure The specific atoms or groups of atoms making up a molecule and the order in which they are connected to one another. Primary structure says nothing about the shape of a molecule, only which atoms the molecule is made of and which atoms are connected to each other and in what order.

Prokaryote An organism that has *prokaryotic* cells.

Prokaryotic The quality of not having a cell nucleus.

Protein A polymer of amino acids.

Purine A class of nucleotide base with *two* rings: one is a five-atom ring with two nitrogen atoms, and the other a six-atom pyrimidine ring. The purine's two rings are covalently attached to one another in a way such that they share two carbon atoms.

Pyrimidine A class of nucleotide base with an aromatic ring containing four carbon and two nitrogen atoms.

Pyrophosphate bonds The covalent bonds between adjacent, covalently linked, phosphate groups.

Quaternary structure A molecule structure in which two or more *tertiary structures* attach or associate with one another to form an even larger molecule or complex.

Ramachandran diagram A plot of peptide bond angles showing contours of which angles fall within certain energy ranges.

Random coil A conformation in which a polymer has no specific secondary structure.

Reference state A state of a system chosen to be used to define all other states of the system. The other states of the system are defined in terms of the difference in energy (or some other quantity) between them and the reference state. Often the *ground state* is chosen as the reference state.

Reflection Light (electromagnetic radiation) bouncing off a surface.

Refraction The bending of light (electromagnetic radiation) while passing through a surface.

Regulation The control of a biological process by a cell or organism, usually through some kind of a feedback mechanism.

Residue Each of the units or smaller molecules making up a polymer.

Ribose A five-carbon simple sugar (carbohydrate) commonly found in nucleotides.

Ribosome A complex of structural proteins, enzymes, and ribosomal RNA, a basic structure involved in protein synthesis.

Right-handed helix A helix that turns clockwise as you move along its length.

RNA (ribonucleic acid) A nucleic acid whose nucleotides contain ribose. See also *mRNA* (messenger RNA), *tRNA* (transfer RNA), and *rRNA* (ribosomal RNA).

RNA polymerase An enzyme that catalyzes the biochemical reaction to covalently link ribonucleotides together to form RNA.

Saccharide Another name for carbohydrate. (See also *monosaccharide, disaccharide, oligosaccharide,* and *polysaccharide.*)

Salt bridge An ionic attraction between a positive and negative ion, usually between two parts of the same molecule (such as a protein) or between two molecules within a complex.

Saponifiable lipid A lipid that is a polymer or that contains a polymer as a major portion of its structure. See also *nonpolymeric lipid*.

Saturated fatty acid A fatty acid with *no* double bonds in the hydrocarbon chain.

Scanning electron microscopy (SEM) A technique in which a beam of electrons is gradually scanned across the surface of the specimen to provide a highly magnified image of the sample.

Second law of thermodynamics For any process, the sum of the entropy of the system plus the entropy of its surroundings will only increase or remain constant. If the process occurs in a *closed system*, then the entropy of the closed system will only increase or remain constant.

Secondary active transport An active transport where the energy needed for the transport comes from a simultaneous *passive transport*.

Secondary amine An amine covalently connected to only one hydrogen atom and with two covalent bonds to the rest of the molecule.

Secondary structure The simple, three-dimensional structure of a molecule. Some large molecules may contain more than one secondary structure (see also *tertiary structure*).

Sedimentation The motion of particles in a fluid under the application of a force.

Selectively permeable Having a quality such that some things can pass through while others cannot.

Self-assembly The forming of molecular structures through folding and/or through association with other molecules without need for energy input, catalysis, or other molecules that are not part of the structure.

Side chain In an amino acid, the atom or group of atoms, other than the amine and the carboxyl, that is attached to the *alpha carbon*.

Single bond A covalent bond in which a pair of electrons is shared between two atoms.

Sodium-calcium exchanger A membrane transport protein that moves three Na^+ ions into the cell for every Ca^{2+} ion that it moves out of the cell.

Specificity A property of some proteins and other molecules that only one certain type of molecule will bind strongly to it.

Spectroscopy Any of a variety of techniques that measure and present data according to a spectrum, that is across a wide range of some variable such as frequency, wavelength, energy, or mass. See also *EM spectroscopy*.

Spectrum An arrangement of data or component parts of some phenomenon according to frequency, wavelength, energy, or mass.

Stacking interaction An attractive force between two aromatic rings, as the result of synchronized fluctuations in their electron clouds that can occur when the aromatic rings lie in parallel planes in a stacked orientation.

Statistical mechanics The application of probability and statistics to large populations of molecules, in order to provide a molecular interpretation of experimentally measurable quantities.

Steric interactions Repulsive forces that arise from atoms coming close enough for their electron clouds to begin to penetrate one another. Also called *steric repulsions*.

Stirling's approximation An approximation for $n!$ Stirling's approximation is very close to the actual value when $n > 60$. The formula for Stirling approximation is $n! \approx \sqrt{2\pi n}(n/e)^n$.

Subunit Any of the individual molecules in a *complex* or *quaternary structure*.

Subunit binding See *molecular binding*.

Supercoil A single turn of a superhelix, or sometimes may be synonymous with *superhelix* (as a noun). Supercoil (as a verb) can mean the process of forming a superhelix or of being in a superhelical state (see *supercoiling*).

Supercoiling The process of forming a superhelix or of being in the conformational state of a superhelix.

Superhelix A *tertiary structure* in which the axis of a helix is itself curved into the shape of a helix.

Surroundings The part of the universe that does not include the *system* as defined for discussion; the rest of the universe, relative to the system. See also *system*.

Symport Cotransport where the different types of transported molecules or ions are all being transported in the same direction.

System Any part of the universe that has been defined as a separate entity for purposes of discussion or experimentation; the part of the universe that we are interested in. See also *surroundings*.

Targeted delivery system A system that uses lipid vesicles, or other microscopic objects, to deliver drugs to very specific locations, or even specific cells, within the body.

Temperature-scanning absorbance spectroscopy A type of absorbance spectroscopy in which the absorbance at one or more specific wavelengths is measured as a function of temperature.

Tertiary amine An amine group where the nitrogen has three connections to the rest of the molecule, and no hydrogen atoms directly connected to the amine nitrogen.

Tertiary structure A three-dimensional molecular structure that is the result of one or more *secondary structures* (within a single molecule) rotating around covalent bonds in such a way as to bend, twist, or fold in relation to one another.

Thermodynamics The study of energy and how it operates in the physical universe.

Thylakoid A disklike structure, in chloroplasts, where photosynthesis takes place.

Topoisomer Individual DNA molecules that are identical except for their linking number.

Topoisomerase Any of a class of enzymes that change the level of supercoiling in DNA by changing the linking number,

Topological invariant A property or quantity that does not change under continuous deformations of shape.

Topology The branch of geometry and mathematics that deals with objects being bent or deformed in a continuous manner.

Transcription The process of synthesizing mRNA from DNA.

Translation The process of protein synthesis, starting with the sequence of nucleotides in mRNA and translating the mRNA sequence into the corresponding sequence of amino acids in a polypeptide chain.

Transmembrane protein A *membrane protein* that spans the entire thickness of the membrane.

Transmission electron microscopy (TEM) A technique in which a beam of electrons is passed through a very thinly sliced sample to provide a highly magnified image of the sample.

Transport protein See *membrane transport protein*.

Triple bond A covalent bond in which three pairs of electrons are shared between two atoms.

tRNA (transfer RNA) A type of RNA that transfers each amino acid to the correct location in the growing polypeptide chain.

Tubulin A type of protein molecule that can be polymerized to form *microtubules*.

Turbulent flow Fluid flow in which there are eddies: small currents that flow backward, or in a partial circular motion, opposite the direction of the overall fluid flow.

Twist (Tw) The number of times one curve wraps around another curve along the entire length of both curves.

Type I topoisomerase A topoisomerase that operates by a mechanism that breaks only one strand at a time and changes the linking number one at a time.

Type II topoisomerase A topoisomerase that operates by a mechanism that breaks both strands of DNA and changes the linking number two at a time.

Ultracentrifuge A *centrifuge* specially designed to spin at an extremely high rate of speed.

Unsaturated fatty acid A *fatty acid* with at least one double bond in the hydrocarbon chain.

Vacuole A giant vesicle. Vacuoles are often made from many vesicles that have fused together.

Valence electrons The outermost electrons of an atom that can participate in chemical bonding.

van der Waals force In this book, all molecular forces that are not covalent bonds and not crystalline ionic bonds. Some books may use *van der Waals force* to refer only to forces that result from charge fluctuations (dispersion forces) or to some other subset of noncovalent and nonionic bonds.

Vesicle A small, approximately spherical, lipid bilayer container; also called *lipid vesicle* and *liposome*.

Vesicular transport Transport across a membrane via liposomes that contain the molecules being transported in their hollow center; see also *endocytosis* and *exocytosis*.

Voltage clamp A technique used in electrophysiology to measure and characterize electric currents across cell membranes by holding the voltage constant.

Voltage gated ion channel An *ion channel* that opens or closes depending on the voltage across the membrane.

Writhe (Wr) A measure of how much a particular curve does not lie in a plane or on the surface of a sphere (but rises above or below it). $Wr = Lk - Tw$.

X-ray crystallography A technique for determining the relative positions of atoms within a crystal.

Zwitterion An ionized molecule that has both positive and negative charges on it, but a net charge of zero.

Index

Note: Page numbers followed by *f* denote figures; page numbers followed by *t* denote tables.